Unnatural Selection?

Books by Edward Yoxen

The Gene Business

Unnatural Selection?

Unnatural Selection?

Coming to terms with the new genetics

Edward Yoxen

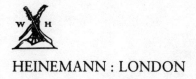

HEINEMANN : LONDON

William Heinemann Ltd
10 Upper Grosvenor Street, London W1X 9PA

LONDON MELBOURNE JOHANNESBURG AUCKLAND

First published in Great Britain 1986
Copyright © Edward Yoxen 1986
ISBN 0 434 89390 0

Photoset by Deltatype Ltd., Ellesmere Port
Printed and bound in Great Britain by
Mackays Ltd., Chatham

Contents

Acknowledgments

Making one's acknowledgments for help received is like making a short speech of introduction. Although you may rehearse it several times, there is always the thought in your mind that someone will get forgotten. In writing a book like this one, which draws upon literature in several fields, I have necessarily accumulated a large number of intellectual debts, as people have referred me to new work, corrected my mistakes, challenged my interpretation and offered encouragement in a variety of ways. It is a pleasure to thank Mary Barton, Valerie Beral, Mark Dixon, Michael Hull, Margaret Jackson, Marie Johnston, William Mosher, Simone Novaes, Philip Parker, Dudley Poston, Mike Rayner, James Schlesselman, Jon Turney and the Women's Reproductive Rights Centre for their generosity in answering my queries and sending me material, often prior to publication and without any prior acquaintance. Various people have also read sections of the manuscript. I know that it was much improved by their comments upon it, both in specific places and in general. I thank Beth Alder, Paul Hodgkin, Becky Holmes, Ramona Koval and Naomi Pfeffer for their friendly criticism and the encouragement which was clear even when our views differed. I am also very grateful to Beverly Hyde, Jonathan Harwood and Ken Green, for the support and help I received from them, as also to my agent, Caradoc King, and my editor at Heinemann, Amanda Conquy, for their work in keeping me going, as the task of writing this book stretched out in time.

Finally there are others, whom I would like to thank, but we have an agreement that this shall not be done here, since actions speak louder than words, whatever writers may think. Nonetheless I hope this oblique reference is sufficient to indicate the nature of this debt and that I am aware of the generosity of which I have been the recipient. It remains to me to prove it.

Edward Yoxen
Manchester
May 1986

Chapter 1

Unnatural selection

In search of the unnatural

There is a great deal of sex in this book, quite enough to shock the faint at heart. Scarcely a page escapes some reference or allusion. But, with the exception of this opening paragraph, there is not even the slightest dalliance with the erotic. Nothing is described explicitly, but none-theless sex is there throughout as an underlying theme, treated to a remarkable number of variations. That being said, this is a book intended to help people cogitate rather than phantasise. It is about reproduction rather than sexual activity. Nonetheless the two are intimately connected, always have been and always will be.

Now, however, matters are getting just a shade more complex, because of the tremendous advances in biological understanding since the 1960s. Sex and reproduction are not divorced, nor, I believe, will they ever be, contrary to Aldous Huxley's vision in *Brave New World*. But these two old partners, linked together for aeons, are now finding ways of loosening their association. Contraception permits sex without reproduction; new reproductive technologies, like in vitro fertilisation, permit reproduction without sex. Or so it is to a first approximation; sexual activity may be transmuted or marginalised, but it never goes away completely. Some of the medical procedures, like artificial insemination, have a residual sexuality, which cannot be totally erased. Indeed it may be deliberately emphasised, to re-introduce old feelings to new practices.

Reproduction, however it is brought about, and once upon a time the options were very few indeed, excites our very most basic hopes and fears. We may talk of it very abstractly, but not far beneath the surface a strong current of emotion is to be found. In that sense the new possibilities do not change anything. The new technologies have old undertones and address familiar needs.

But new possibilities there now are: artificial insemination, in vitro fertilisation, embryo donation, embryo freezing, antenatal diagnosis, and sex selection. Soon genetic engineering in human beings and cloning may be possible. With unrepentant imprecision, I call such things the new genetics.

Such matters were first placed before the public in the 1960s. The growing technical power and social visibility of biological research made the discussion of what were then somewhat distant possibilities more plausible. In the 1960s some of the post-war optimism for science as inherently progressive disappeared. People asked themselves what, if its enormous powers of analysis and control were to be concentrated on genes and reproduction, frightful possibilities might be realised, and with the sincerest intentions. We were, it seemed then, in some danger from our own cleverness. It was only a question of time before the biological time-bomb went off.

In the intervening twenty years a very great deal has changed both scientifically and culturally. A surprising amount that used to be spoken of as possible only at some very indeterminate point in the future has already happened. The first test-tube baby was born in 1978 and the first baby born after having been deep frozen for a time as an embryo was born in 1984. The first attempt at embryo donation was described in 1983, and although it is not often done it is clearly feasible. The first attempt to correct a genetic defect in a human patient, a form of genetic engineering, was made in 1980. It is said to have failed, but better prepared plans to try again are now being discussed. Genetic engineering in mice already works, with dramatic increases in their size. This was reported in 1984. Scientists are now applying this technique to agricultural animals. Identical sheep and cows have been born after dividing one embryo into its constituent cells. Mice with more than two genetic parents have been created by combining embryos. One could almost say that only the genetic engineering of human embryos and human cloning remain as technical feats still to be pulled off.

On the cultural level much of the scepticism about science and interest in alternatives has disappeared in the harsher climate engendered by recession and policies for economic reconstruction. The heady optimism that science could easily be steered in new

2

directions or turned off or transmuted into more controllable forms has turned out to be misguided. Research is both too important to economic vitality and too resistant to simple analysis to make that possible. It is remarkably resistant to change and now increasingly tightly coupled to industrial objectives. For many it 'works' and that is that. Some, indeed, see scientific entrepreneurship as the means to both personal and collective salvation, after the self-indulgence of the 1960s and the economic collapse of the 1970s. Leaner, fitter companies need more mobile, less fastidious, more commercially oriented scientists, able to link their biological knowledge to their employer's business plan. Against this background, questioning what science is about and where biology is taking us might seem to be a marginal, if not entirely ludicrous, activity.

Thankfully that is not the case. In fact concern with the social implications of biological research has never been stronger, helped by the spreading realisation that many things have moved from the realm of speculation to become matters of routine in a very short space of time. The sense that there are a number of issues to resolve, centred around novel forms of procreation and genetic research, is widespread. Many people are simply amazed by the speed of developments and find it hard to know what is the appropriate response. Very few are completely indifferent. Now that we have finally got down to thinking about new reproductive options, the consensus is exceedingly precarious, and public policy is constantly shifted by upwellings of buried sentiment. In this book I am trying to speak directly to that unstable sense of concern.

Some people clearly find all such scientific developments profoundly shocking and offensive. They think, talk and write of test-tube babies, antenatal diagnosis and genetic engineering as 'unnatural' and something to be shunned with the utmost vigour, their moral antennae trembling violently. For many more, whose ideas of nature are more plastic and imaginative, these new possibilities are very unsettling, but not immediately abhorrent. For them, the unfolding options are just a little too subversive for comfort. They demand too many adjustments and re-definitions too quickly. I share some of these anxieties: writing this book is in part a personal attempt to deal with them. But I do not share the view that the new reproductive technologies have broken a fundamental taboo or ripped a hole in the moral fabric of our society. I think we need to think through what is likely to happen and what moral, political and cultural responses are appropriate. Mere distaste or outrage are simply not adequate. Analysis and calculation need to complement moral fervour. Hence the subtitle to this book, *Coming to terms with the new genetics*.

This cogitation ought to lead somewhere. I hope it leads to an increased sense of confidence that very novel and undeniably disturbing possibilities can be handled. We must ask ourselves how the new reproductive possibilities are to be assessed. What, for example, do they resemble? What emotions would their application call forth? Which values would be threatened and which enhanced? How could these new options be developed so as to increase people's confidence in their own abilities to control their lives. I retain my faith in the progressive potential of science. I believe it can be used to liberate as well as to oppress. The problem is to create the kind of egalitarian society in which these beneficent uses of science are the only ones that are thinkable.

The idealistic notion that we can generate the power to control science, appraise its implicatons before they are upon us, and shape them to our own ends, must first survive a rather crude but all too effective form of attack. It is often said that the applications of the new genetics go 'against nature'. On this view, which I believe to be profoundly reactionary, the new possibilities are unnatural and as such not to be contemplated. I have acknowledged that such sentiments exist with the main title to this book, *Unnatural Selection?* with its evolutionary undertones. But note that I have added a question mark. I can best represent the attitude at work here with a series of rhetorical questions. Are we about to leave nature behind? Are we abandoning natural constraints, only to voyage into chaos? Are we about to acquire powers of selection that transcend what is merely natural, taking upon ourselves awesome responsibilities, that we shall mismanage? To all this my reply is: what does such language mean? The symbolism of what nature permits or demands is exceedingly potent, in any culture, ours included. To call something unnatural is to utter a dire warning, and to offer a moral prediction, that the activity, whatever it is, will end in tears. But even so, do we actually gain anything by speaking this way?

The new genetics certainly deserves such serious treatment. Concern is not out of place. We need to invest time and energy in evaluating research that stands to re-order so much about our lives, our sexual relations, our basic satisfactions, our kinship, our sense of meaning to life and what we have on our conscience. Taken as a whole, it is not mere technicality like a new internal combustion engine or a new way of getting oil out of the ground. It concerns how, whether, when and with whom members of our species reproduce.

But the language of nature and its opposite, of acts natural and unnatural, is a very tricky one to use. Often it is used collusively, presuming or implying a shared agreement about common standards where none in fact exists. We all *know* that contraception, or adultery, or taking drugs, or practising euthanasia is unnatural. Well, don't we? End of argument. This is knee-jerk evaluation, hurling into the social waste-paper basket something unusual that elicits an unthinking response. An opportunity for reflection and argument is lost. The unnatural stands condemned from the start and can only be the subject of open-and-shut cases. We learn nothing from such summary justice. This dismissive attitude is worse than useless in the face of medical and scientific novelty, where change, if not progress, is constant. Science is, by definition, the transformation of nature.

Come to think of it, what is nature? What is this supposed standard of reference for all that is socially normal and healthy? How and when does anything become unnatural, rather than simply unusual, or untried or unpromising? These are weighty issues which I cannot settle in an introduction, but my main and obvious point is that the nature of nature is ambiguous. When we use the word it is far from obvious what we really mean. Does nature mean what animals do? Or what would happen on the planet if there were no human beings, a recreation of the Garden of Eden? Or what 'noble savages do? Or human activity without recourse to technology? All these are rather different, but are quite acceptable meanings of nature. Many animals are instinctive predators, abandon weak members of their species and lead exceedingly pragmatic lives. If this is how 'nature' operates, should we do likewise? Many primitive tribes are deeply xenophobic, worship a variety of gods and spirits, and some eat their elderly relatives when they die. If this is 'natural', ought we not to emulate it? Life without technology, without fetal heart-monitors, without food additives, and without heart-lung machines, may give us natural birth, natural food and natural death, which would be good for us, but carried to extremes on the way back to nature, it would also give us natural hunger and natural disease, about which I personally have some reservations. So a word we love to use, which has such powerful connotations, seems to fall apart in our hands when we examine it. This is partly real, partly illusion.

In fact nature is a construction. Ironically, the term denotes artefact. We select bits of the external world, and call them nature. We put together models of how we think the world is and of how we think it should be seen. Each culture does it differently. What we call 'natural' is a convention, derived from a way of seeing and classifying things that appears stable but is in fact transient. Even though the order of things changes, the whole impression of constancy is very powerful and important. As a framework it is important, if not vital, to any society, as a moral guide. It expresses basic distinctions and rules. But these ideas are not and cannot be external and unchanging. They codify the received wisdom.

Even though we really ought not to conclude that whatever is natural is good, and whatever is unnatural is bad, we constantly tend to do this. Undertones evade logical constraints, and an approving or a disapproving attitude is transmitted in our very choice of words. So is it useful to speak of natural and unnatural reproduction, given that all these possibilities are just variations on a basic form, and all rely on 'natural' processes like fertilisation, development and birth? They do not depart very far from the basic, historic conventions. They combine natural elements in new ways. That is just the point. We had a set of conventions for describing and practising reproduction – or conceiving conception, if you will – we now need another, because new ways of facilitating human reproduction are possible.

Under the influence of applied biological research the utility of our existing linguistic and moral conventions is breaking down. We need to change them. We ought to stop speaking of new forms of reproduction as 'unnatural', since that only creates anxiety to no useful purpose. Some might think this represents a wilful abandonment of any standards. For if we forswear this form of condemnation, then surely anything is permitted? Far from it: all we have given up is a rather tendentious way of speaking. We still have to decide what is right and wrong. Taking such decisions is both harder, because we actually have to think about it, and easier, because the fact that something is new and unusual is no longer a distraction. We must expect rather than shun novelty.

At the same time the symbolism of nature is too powerful to relinquish entirely. It can, for example, also be a source of reassurance and support, if we need to emphasise the goodness and common sense in some new arrangement. What we must not do is to use this language unthinkingly and uncritically. It is a moral resource to be tapped with care.

New science, new language

Quite the most remarkable and unexpected cultural effect of the new genetics is its impact on language. One after another all kinds of familiar words no longer 'work' in situations where once they were quite appropriate, where their use seemed easy and, dare I say it, natural. This is very far from being a trivial thing, and paying attention to it is not an academic pastime. It is an immensely valuable sign that some kind of cultural adjustment is going on: the social processing of new ideas manifest in speech. The new language of description and debate is worth listening to, literally, because the hesitations, undertones, allusions and attempts to purify new terms of past associations can tell us a great deal about underlying hopes and fears. Our language bears the weight of new problems and dilemmas and it shows. A fine example of this involves the terms we use for parenthood.

In English the historic convention is that 'mother' means female genetic parent and the woman who also gives birth to the child. If she continues to look after the child, she takes on the additional social role of a mother. 'Father' likewise is taken to mean male genetic parent, although that may represent the extent of his contribution. Curiously, when we use these words as verbs, 'fathering' usually means begetting or procreating – making a genetic contribution – whereas 'mothering' suggests nurturing and caring, with no particular implications of a genetic connection or of an active and necessary role in pregnancy. The asymmetry is clearly sexist. Parenting, too, has connotations of care rather than creation.

Similarly when we dip into Latin, paternity and maternity tend to have different meanings. The former usually stands for the genetic relation between man and child, as in paternity testing, whereas the latter refers to giving birth and early child care, as in maternity hospital and maternity leave. Whilst genetic paternity is not infrequently questioned, maternal genetic relations are almost always assumed to be self-evident.

Our current ideas of parenthood in Western society acknowledge a distinction between its biological and social aspects. But we take the former to be primary, as constituting a factual bedrock of genetic relations on which subsequent relationships may be built. At the same time responsibility for children is transferable, to some extent. If one parent dies or withdraws and another adult enters the household we speak of a step-mother or step-father.

Adoption is similar and here the transfer of parental rights and obligations is done by the courts, in some cases conferring social parenthood on people who have no genetic connection with the child they wish to regard as their own. But here already the simple device of a prefix like 'step' is not enough for some reason. We often talk of the 'natural' parents or the 'real' parents when we mean the people who have given up their child for adoption. The words natural and real are rather unfortunate in this context, even though it often makes some sense to indicate that a particular family has been created or extended through adoption. 'Real' parents implies that adoptive parenting is unreal or imaginary, when it is perfectly real and depends upon relations of substance. Similarly natural parenthood invites the thought that adoption is unnatural, which is not a helpful thing to say. 'Original' parents would be both honest and clear. Our language then already shows signs of strain here. It is of course possible for people to combine these roles and to be natural, adoptive and step-parents all at the same time. They could even be father- or mother-figures as well. Such is the change in family structure today, that people sometimes use the remarkable term 'blended families' to denote those in which several different kinds of parent-child relations exist simultaneously.

Suppose now that instead of being adopted after birth children come into a family in the following way. A married couple discover to their dismay that they are both infertile. The man produces no sperm, and can therefore never be the genetic parent of a child, at least given the present state of biology. The woman has ovaries but they do not function properly. She has a womb, but is advised that for medical reasons a pregnancy would kill her. She too is infertile. Despite these problems they are desperate to have children, and are not keen on adoption. However, her sister is herself married and has several children, and she offers to act as a surrogate mother, that is to bear a child for them and then hand it over after birth. What is proposed is that an embryo created by fertilisation outside the body and donated by another couple should be transferred to the womb of the infertile woman's sister, who will then carry the developing fetus and eventually give birth. The child is then to be handed over to the surrogate's sister and her husband, and they become parents.

This kind of thing makes the head spin, even though it is no more complicated than the multiple adulteries and pregnancies of a reasonably salacious soap opera. We must also ignore the question of why anybody would want to go through this complex procedure rather than apply to adopt a child and we must assume that all the doctors involved are quite happy with what is proposed. In this example there is a new baby, with three possible 'fathers' and three possible 'mothers'. Clearly we are here in a rather novel world and our language is not subtle enough to register all the distinctions. Indeed the adjectives 'social' and 'biological' have lost some of their power to help as well, because two women have an undeniably biological connection with the child, though neither wants to be the eventual social mother.

I could easily generate a whole set of ever more complex examples from this one case. With the addition of deaths during pregnancy and divorce under the strain of such arrangements, or the unexpected appearance of twins, or the sisters themselves being identical twins, cases like this one can be made even more complicated. Such elaboration quickly becomes totally confusing and frankly tedious. But the exercise illuminates two important preliminary points. Firstly, technical advance in biology has destroyed forever the simplicity of speech with which we have grown up. The example above could not be realised without recent work in reproductive biology. It is now entirely possible. Secondly, it always used to seem that the responsibility and obligations of parenthood rested quite straightforwardly on biology. In fact matters were never that simple, as the myth of King Solomon's judgment about motherhood suggests. But now more than ever it is clear that such ideas about who belongs to whom depend on social agreements about which biological facts are to count. We now have to define motherhood and fatherhood rather more precisely, given that different aspects of one role can be played by different people. Biology alone cannot decide matters; we have to take a social decision about which factors we wish to consider significant. Reference to some natural standard cannot be a helpful guide in some situations which we can now create.

Connotations

Our language then can scarcely cope with the burden placed on it by scientific ingenuity. New technology requires new terminology. But there is a twofold problem here. On the one hand, old words cannot quite match up to the new demands placed on them; on the other, every linguistic innovation has its problems, because of the connotations of past usage. Words trail their pasts behind them; or if they are utterly new they are too bright and shiny to be comfortable to use. Their very unfamiliarity is itself disturbing, at least for a time.

For example, consider 'artificial insemination'. The latter word is a very clinical term. It has no place in ordinary speech, except perhaps amongst farmers, whereas fertilisation and impregnation do. It specifies the activity which makes it different from the conventional mode of procreation. But the very precision of reference is itself slightly alienating. Add to that the connotations of 'artificial' – not natural, done by artifice, simulated or using man-made [sic] materials. The implications are unmistakable; this is something odd, something second-best, something that is being made to work.

There are of course alternative terms, some of them hilarious, some pompous, all revealing. In the nineteenth century artificial in-semination was sometimes called artificial impregnation, which shifts the semantic centre of gravity from the introduction of semen to the reproductive tract to the initiation of pregnancy itself. It moves the point of reference along the chain of causes and effects. It refers more to the outcome than the process, and as such it covers more events with a veneer of artificiality.

Another term in use was 'aetherial copulaton', invoking spirit action, and introducing thereby a fictional sexual act to 'explain' the pregnancy, as a result of sex, but not with a person. To modern sensibility this just seems to make matters worse and to reintroduce confusing feelings in an agony of embarrassment. Such is our relation to the spirit world in more secular times that it is hard to see this description being useful today. The term 'therapeutic donor in-semination' is sometimes used nowadays. In its medicalising way this is every bit as bad, because no one is being treated for an illness, so to call the action therapeutic can only be misleading. I come back to this point again in Chapter 2. I follow the slightly evasive practice of using the acronym, AID. This of course solves nothing, because even initials have connotations. Thus AID sounds like AIDS, and looks a little like AI (artificial intelligence).

Or we could speak of 'fuckless babies', as several feminist writers have suggested. Their intention is to remind us that women can have babies without going through the often brutal, loveless experience of being fucked. Whilst I find the deliberate coarseness of the term appealing, and a refreshing change from classical borrowings, it is too close to luckless or hapless for comfort; and, to turn the argument around, the absence of sex in procreation does not imply the absence of love.

Then consider the phrase 'surrogate mother', which we often compress to the surrogate. This is a curious word. Why we have

10

suddenly decided to use it is a mystery, when 'substitute' or even 'temporary' would do instead. Sometimes people talk of the 'carrying mother', i.e. the woman who is pregnant, and carries the child to term. Perhaps surrogate appeals because it can stand alone, without any complicating reference to motherhood? But then so can substitute and so can 'temp'. Whatever is the case, English usage indicates a desire for a special term to name a special role, one that we like to think is rather new. It is surely no accident that the other contemporary usage of the term surrogate is in sex therapy, where someone, usually a woman, acts as the paid sexual partner when intercourse or physical intimacy has become a massive source of anxiety. Surrogacy implies activity on the frontiers of the permissible.

But in surrogate motherhood what is being substituted for? Who is the surrogate? Is she the woman who is pregnant, who stands in for the carrying mother? Or is it the woman who is given (back) the baby, and who herself stands in for the care of a child born to someone else. These are not just linguistic points. They relate to the rights and feelings of the women involved, their self-image and their sense of indebtedness. Almost as problematic is the terminology that is evolving to describe the other parties involved, who are often called 'the commissioning couple', a couple being presumed most of the time. Are they commissioning a baby, as one commissions a painting or a string quartet or plans for a house extension? They place a contract for a service and something comes into existence for them, so perhaps the term is appropriate, even desirable to underline the commercial nature of the activity? But we do not use the same language in adoption, even though, leaving the money aside, the relations between people are not all that different. To put the point another way, we tend not to refer to the people involved in surrogacy as adoptive parents, the term used in classical adoption, even though that is what they are.

Or consider the terms 'test-tube baby' and 'in vitro fertilisation'. Fifty years ago the term 'test-tube baby' was used to denote children born after artificial insemination with the husband's semen. It now refers to babies conceived outside the body by in vitro fertilisation. I talk of test-tube babies up to this point in the book and then stop because I dislike the phrase. For one thing it suggests that the baby is somehow different, whereas the differences exist only at a much earlier point in the life-cycle. Fertilisation occurs outside the body, under laboratory conditions, but then the whole of pregnancy and birth proceeds as normal. We might as well speak of test-tube men and women. That would be as accurate. I suspect we talk of babies here because that is the most popular term for a human being before birth. Embryo is already a technical term, but I am going to use it, because I think the babification of the whole of development and pregnancy is wrong. Embryos are not babies; in the human species only a small proportion of the former become the latter. Since the law and the moral code shared by many acknowledges this fact, our language should really do the same. So, in this book, no more test-tube babies, even though the shorthand is very convenient.

The term 'in vitro fertilisation' has its problems too. It is robustly technical, proclaiming its status by being partly Latin. It means fertilisation outside the body, in laboratory glassware, although the glass can be non-toxic plastic. There are alternative terms one could use, one of which I almost picked for this book. That is 'external human fertilisation', which Clifford Grobstein popularised in his book *From Chance to Purpose*. At least it is not Latin, and therefore sounds less forbidding. But it is also slightly vague; external to what, after all? I could also have used the term 'vital initiation of pregnancy', which one American specialist prefers, but VIP is downright confusing. In 1974 the *British Medical Journal* used the term 'embryo transplants', which is technically correct, but amongst other things, implies a great deal of heroic surgery, which is very misleading. So I rejected that term, although it is interesting that it was thought accurate in the mid 1970s. Artificial fertilisation is also used, but does not seem right for two reasons, firstly, it has the connotations of being unnatural, which I have already discussed, and secondly, there is just a suggestion that this is fertilisation brought about using man-made materials, as is the case when you spread artificial fertiliser on the fields, so this term too is best abandoned. So one comes back in the end to in vitro fertilisation, and its acronym, ivf.

Then there are all the terms that we are now starting to use to describe the application of genetic knowledge in medicine, genetic screening, antenatal diagnosis of genetic disease, carrier, gene therapy, and, of course, the all-embracing 'genetic engineering'. Quite when and where the notion of genetic engineering came from I do not know, but the reference to mechanical intervention is one we make rather selectively. We do not talk of reconstructive surgery of the head as facial or cranial engineering, although a considerable amount of mechanical knowledge and projective geometry is involved. We do not call the management of renal failure 'kidney engineering', even though a great deal of engineering science has gone into the development of dialysis machines. Given that the correction of human genetic disorders is likely in practical terms to resemble existing bone marrow transplants, which we do not call 'stem cell engineering', our term for this application of genetic knowledge stands out somewhat. I think the reason once again is an attempt to indicate difference through a deliberate contrast. It is like the phrase 'reproductive technology', a phrase now in frequent use, particularly by feminist writers, to refer to much in this book, but also to the technologies of contraception, antenatal care and the monitoring of labour, over which many women feel they have no control. It is these undertones of invasion and powerlessness that give an otherwise rather vague term its strength. The two words are intended to work against one another, to suggest something anomalous and troubling. So too with genetic engineering; it functions as a term just because it is slightly paradoxical.

In recent years another term has appeared as a replacement, that sounds more medical and less mechanical. This is gene therapy, which could be therapy with genes or therapy of the genes. As we shall see, it will be both. I prefer this term and I shall use it, when the kinds of intervention intended are therapeutic, because I see no point in maintaining a rather loose description for something so close to realisation. Making gene therapy out to be something bizarre is no longer helpful. But not all kinds of genetic intervention are therapeutic. There are some things that are still best called, for rhetorical purposes, 'genetic engineering'.

Connotations dog our footsteps, as we create words for our innovations. This is not a problem we can evade. There is no language utterly free of undertone, and it is folly to seek one. Rather as we appraise the new possibilities opening up before us we should remain thoughtful about all we are saying with the words that we use, before their use becomes entirely natural.

The method of analogy

It may seem a little odd to begin with so much about language, yet the new genetics or reproductive technology is as much a cultural as a scientific phenomenon. People are moved by a sense of what is possible, and not by direct contact or experience, except in a few cases. Most of us are appraising things secondhand, from the outside, by hearsay and reading others' reports. The rate of diffusion is so rapid that within a few years many of the possibilities discussed below will have impinged on the experience and daily lives of many families in the land. I stress the family context, since it is likely to be through relatives who have had antenatal diagnosis, or children who have been screened at school to see if they carry the gene for a particular disease, or friends whose frustrated desire to have children has led them to artificial insemination or in vitro fertilisation, that many of us will come to appreciate the personal import of the new possibilities. At that point the dilemmas described in public discourse will suddenly come closer to home.

It is vital therefore that as many people as possible start thinking now about what this might mean to shape a considered response. It is not that these new technologies are likely to be imposed willy-nilly upon us, or that we must awake from a moral slumber, in order to reject them wholesale, but it lies within our power to influence the form of these medical innovations, to set guidelines and limits for them, and to require the fullest public discussion of all their implications. We cannot do this if we continue to feel that these things are alien, that they lie beyond our comprehension, that they will happen whatever we do or that few people have any competence to comment upon them. We need an idiom for thinking about imminent possibilities that gives us confidence to hold them in our imagination, where we can turn them around and examine them from all sides. We need to work out what they might feel like. In an earlier book I called this activity of anticipation 'culturing futures'.

To help this process accelerate, two things are necessary: one is a set of terms that are clear and accessible, the allusiveness and redolence of which are being recognised and valued; the other is a set of concrete examples to help us perceive both what would be possible and to decide what would be good. Accordingly in each chapter I describe some hypothetical cases. These always have some analogy with familiar arrangements today, and the relation of analogy is absolutely basic to my purpose. Some of these examples form a sequence of deepening complexity. My object is to build bridges between what we know to happen at present and what could occur in the future. I call this 'the method of analogy'. This is not to say that just because something happens now it will be familiar to everyone, and even less that is is morally acceptable, but simply that the form of new reproductive possibilities is never utterly new. They will always resemble present arrangements to some extent. I would even say this of cloning, which might seem the epitome of unnatural reproduction, frequent though it is in nature. If we were trying to decide what we felt

14

about human cloning, we could easily think of analogous cases, i.e. cases with *some* similarity: single-parenthood, children of the same sex as famous fathers or mothers, children in polygamous households with a common parent, and so on. Analogy is heuristic; it helps us to think, and thus to feel.

My underlying purpose is to popularise an idiom for non-expert debate of the new possibilities, and to bolster a general sense of confidence that we have the imagination and the insight to decide what we want from biology. My examples then are intended both to display the moral and psychological complexity of new options and to show that we do have ways of thinking about them. We have no need to recoil in shock: we have no reason to hand over their appraisal to experts in the law, in medicine or moral philosophy, even though their contributions are, for the most part, a necessary part of public debate. Instead all of us can and should participate in the extended discussion of how the new genetics is to work for us, in a society in which personal autonomy and the development of individual responsibility are seen as primary values. This for me is the primary issue in relation to science. I am not unconcerned about setting limits to inquiry. I believe that there are taboos on some possible subjects for investigation and that some scientific procedures are unacceptable. Research needs to be licensed and controlled. But I am more concerned that people should feel that science confers upon them the power to live their lives as they see fit. The question is not whether a new procedure is unnatural, but whether it leads to a sense of freedom. My dream is of a society that has so organised its relation to biological expertise that people may say to themselves, I fully understand this step I am taking, I have chosen it freely, having sought advice and considered the options available, and I recognise my responsibilities to myself and to others in what I do here. Modern biology has the potential to empower and embolden people in this way. Increased understanding of the physical world ought to dispel some of the uncertainty and make us more confident to act.

Research today deposits more and more technical expertise in the hands of scientists and doctors, who countenance and just occasionally invite discussion of the implications of this accumulating technical facility. But the tide of research quickly flows on, around such islands of reflection and analysis. And it is rare indeed for the question to be asked as to whether a particular development will make it easier for people themselves to exert greater control over their lives, rather than allowing others to exercise that control on their behalf. That question

15

of how to create autonomy and responsibility defines the most important theme in this book. It is the best reason I have for putting pen to paper. It shapes the approach that I take to each of the topics under consideration.

We begin with something that does not depend on state-of-the-art research. It is very simple to undertake. This is artificial insemination with donor semen. It is now almost as common as adoption in countries like Britain. It raises as clearly as any topic the social issue of personal autonomy and confidence. One could certainly describe it as unnatural, if one really felt the need. Or one may ask, as I do, how it could be organised so as to build people's confidence and esteem, if they have recourse to this procedure, and to help them recognise and handle the psychological problems involved.

Then we move on to in vitro fertilisation (ivf). I devote considerable space to analysing the case for ivf. I also try to make clear how many elaborations of the basic technique there could be. I also discuss, partly because it is such a topical issue at the moment, the morality of research with human embryos. But the central question is whether ivf will promote a sense of freedom and self-control. For example, will it help women to feel that the control of their fertility lies within their power? There is no simple answer to this question, and in any case it would be stupid for me, as a man, to pronounce upon it definitively. I can see both positive and negative developments. However, as a technology I believe it has occasioned a valuable debate about the causes of infertility, the risks of experimental medicine and the desire to be a parent. Ivf could be frozen into a very familiar technological form, under expert control, as a means to alleviate what are represented as rather specialised medical problems. Or it could engender a deepening debate about research priorities, the role of the medical profession and personal reproductive autonomy. It is not just that it might help a few couples have the children for which they have longed, but also that it could help all of us to express what we really want from medicine.

From ivf we pass to surrogacy and the predetermination of children's sex. The former has been heavily circumscribed recently in the UK, but is burgeoning in the US. The latter remains in the realm of possibility, but one being brought ever nearer by continuing research. I am critical of both, although not absolutely against either. The question as ever is what would the expansion of these procedures do to people's perception of themselves as responsible agents or as valued individuals. In both cases I believe they could be disastrous for this

16

reason. Whilst I can imagine surrogacy arrangements where every-one's sense of self-worth would be enhanced, I think the opposite is much more likely to occur. Similarly, given the dominant preferences in many countries for male children, at least as the elder offspring, sex predetermination is likely to strike directly at women's self-esteem.

The last part of the book concerns medical genetics and the variety of ways that exist and that will exist to deal with genetic diseases. Already diagnosis in pregnancy of many conditions is possible, and for many people, if they have access to good and sensitive advice this is welcome information on which they readily act. We are also on the brink of a considerable expansion in our ability to determine whether people carry the gene for a particular condition, such that their children may suffer from it. Providing this kind of information in a useful way is much more difficult and I believe we have scarcely begun to consider the problems involved. In particular the question of how people can safeguard their self-esteem and develop their own ideas about responsible reproductive behaviour remains to be thought through. At the same time within the next few years the alleviation of some genetic defects by gene therapy is very likely to become possible. There is certainly a case for allowing this to happen, but again the full social implications of human genetic engineering have yet to be explored. The new genetics could offer us so much that would enhance our freedom; yet it could also make us slaves to a very fixed and restricted conception of what human beings must be.

Chapter 2

Artificial insemination: out of the night

AID as moral opportunity

Artificial insemination is but one way of removing sexual intercourse from the process of procreation. The introduction of sperm into a woman's body – insemination – is done not by ejaculation into the vagina, but with a syringe. In that sense it is artificial. Sperm appear in a woman's body as if intercourse had taken place, but the procedure is designed to underline the fact that it did not.

The procedure is very simple. The woman lies on her back, with her knees bent and the semen is inserted. It is often done in a doctor's surgery, or in a room in a hospital, but it can be done at home. The introduction of the semen is usually done by a doctor, or it may be done by the husband or partner, under the doctor's direction. Or it may be done by one woman for another, without any medical involvement. It can be done by a woman herself, on her own if she wishes. It is then 'self-insemination'. Usually it is done several times in each menstrual cycle, around the time of ovulation, when an egg has been made ready for fertilisation. The object of the exercise is conception.

The statistics of success are such that most people will have to go through several cycles of insemination. The success rate is generally about half that of unprotected intercourse, so it is likely to take longer. Some people conceive within one or two cycles, most within a year; some will give up if unsuccessful, after a while, but a very few persevere for many attempts. Technically simple this may be, but why exchange what we generally think of as the pleasures of intercourse for the detached unerotic procedures of 'artificial' insemination? Why resort to artifice?

One reason is male infertility. Sperm, produced in tissues in the testicles, have to be present in vast numbers in ejaculated semen for one sperm to reach a mature egg in one of the two Fallopian tubes that extend sideways from the top of the womb. Some men produce too few sperm for this to be at all likely and some produce none at all. Sometimes the seminal fluid contains bacteria or antibodies, produced by the man against his own sperm, which inhibit fertilisation. Or the sperm may be insufficiently active or mobile to reach the egg.

Or it may be that the thin mucus lining the vagina or on the outer surface of the cervix makes it much too hard for the sperm to survive. They perish in vast numbers anyway, but the rate of attrition can be too rapid. Doctors often use the most outlandish military metaphors such as 'inadequate cervical invasion' to describe this phenomenon, as if the cervix was a redoubt that had to be taken. In either case the couple attempting to conceive will fail.

Or it may be that the couple who are making use of this procedure are not infertile. They may know that there is a significant risk of their having a child with a hereditary disease. Using sperm from another man can get around that problem. Or the couple may be two women in a Lesbian relationship, who seek to have a child. Or it may be a single woman who has no male sexual partner with whom she wishes to have a child.

There are two basic forms of artificial insemination. One is often called artificial insemination by husband. (AIH) This is a slightly misleading label, because it implies that the couple are married, and they may not be, and 'by husband (or partner)' sounds as if the partner is doing everything himself, whereas it is more likely that a doctor will actually carry out the procedure, using the partner's semen. This is produced by masturbation and is placed inside the cervix, or at the end of the vagina, in the hope that fertilisation will occur.

But rather more often the sperm is not from the marital or sexual partner but is provided by a donor. This is artificial insemination by donor (AID), and here 'by donor' is even more misleading, since he is unlikely to be present and his role is very likely to be minimised as much as possible, and his identity kept a secret. Some people use the phrase 'artificial insemination with donor semen', which is a definite improvement. In this case the genetic parents of any resulting children are the woman receiving the sperm and the donor. Many of the

19

psychological, legal and moral complications of AID stem from this fact.

Thinking about the possible problems with AID, and I shall treat AIH as unproblematic, comes down to thinking about the psychology, morality and legality of donation. For example, what can be given? One can give away sperm, but can one give away paternity? Even if one can do so in law, so that one's rights and responsibilities are reduced to those of a total stranger and non-relative, can one do so psychologically? Equally the receiving of such a gift can be complicated, for both men and women.

For the most part people do not feel able to reveal the fact that their child was made possible by the involvement of another man. Nor in some cases do they find this easy to admit to each other or to themselves; hence in part the recourse to a procedure that requires no sexual connection and which has been stripped of virtually all erotic overtones. But there still remains a genetic connection. This must be dealt with in some way, and given whatever meaning is thought tolerable and appropriate. The whole procedure is then a very carefully crafted mixture of admission and pretence.

Despite these complications, for some people the genetic link between the mother and the AID child represents an advantage over adoption. Or it may be that for one reason or another they have been or would be refused the chance of adoption. AID also means that at least one of the future social parents will have the experience of pregnancy and childbirth; and the insemination itself can be done in ways that enhance feelings of participation by the woman's partner.

How do people find the experience of AID? One might suppose that it generates all kinds of strains within a marriage and between parents and children. It can do. Snowden and Mitchell in their book *The Artificial Family* mention cases where strong resentment has surfaced, or where disappointment with a husband's failure to cope has been a problem.[1] Follow-up studies show that, in general, people adjust easily to the experience, particularly given the pleasure of having a child. It is very likely that as years go by some people's attitudes do change, but how do you disentangle feelings about AID from other effects of time passing? How do you discriminate between a cause of and a pretext for resentment?

It may also be that those who are unlikely to cope with the psychological aspects of AID fail to go through with it. Those who do would then be a self-selected group. In one recent study of Dutch couples in Utrecht, who had sought AID between 1974 and 1979, 20

20

per cent of those on the waiting list changed their minds.[2] But of the 153 couples who became parents in this way, 98 per cent said the decision had been a good one. At the same time very few had told even parents or brothers and sisters about it. 77 per cent said that the child would not be told, although 20 per cent had not decided whether to do so. AID may often be a marital triumph in adversity, but it is still kept very much a secret.

Until thirty years ago the semen had to be freshly produced, each time it was needed. Then it was discovered that a glycerol-egg yolk mixture protects sperm during freezing and thawing. In 1954 Sherman and Bunge announced the first birth born using frozen-thawed donor semen.[3] Freezing makes sperm banking possible. This was discussed in principle in the 1880s and canvassed in the 1930s by the American geneticist, Hermann Muller, in his book, *Out of the Night*. His purpose was explicitly eugenic, to improve the race. Nowadays sperm banks are used particularly by men about to undergo surgery or chemotherapy, as a kind of 'fertility insurance', should they wish to have children in the future. Such sperm is then only for their own use, unless they specify otherwise.

Now people who regard reticence and discretion as the cardinal virtues, a peculiarly British delusion, might ask whether we *really* need to think about these things. Is not AID very rare, very much a private matter and legally and morally trivial? The answers to that are no, yes but so what, and no.

Already AID is almost as common as adoption. If the numbers of children available for adoption continue to fall, and the factors that promote recourse to AID continue to operate, then it will soon come to be the more common experience. It may have done so already, without our being aware of it. Every development to be discussed in this book, except cloning and in vitro fertilisation, makes an increase in the numbers of people seeking AID more likely. A reasonable estimate is that about 250,000 people alive in the US today were conceived after AID. In Britain it might be 50,000 or 0.1 per cent of the population. As a proportion of all live births in the UK it is now about 2.5 per cent.[4] Even if we thought that moral questions only become significant when they affect significant numbers of people, that would be enough to make us think.

Secondly, AID is and should remain something done in privacy. But society rightly interests itself in all kinds of private acts and choices, sometimes to prevent them if possible, sometimes to protect them. I want to bolster the confidence of people having private recourse to

AID. But we cannot achieve that without some public debate. Thirdly, both legally and morally AID is not trivial, as I hope this chapter will make clear. In particular the questions of ensuring the maximum responsible use of AID, disclosure to the offspring involved, and the regulation of sperm banks turn out to be quite complex. Yet permitting AID will not shake our world to its foundations.

When we examine AID we find immediately an obvious tension between procedures and policies that rely on expert control of reproduction and those that set out to limit or remove any such control, placing the power of decision in the hands of lay-people. The value of the former is that it tends to improve technical standards, though there is no guarantee of that. The value of the latter is that it builds people's confidence in their capacity to control their own lives. If one believes that procreation should only take place within the holy state of matrimony, then AID is not acceptable. But such a view undervalues human adaptability, imagination and generosity of spirit. AID is not going to go away or be rendered obsolete; nor is it likely ever to be entirely straightforward. Our objective, then, should be to minimise the complications and to help people feel that they can handle those that remain. It is far better to treat the continuing expansion of AID as an opportunity to broaden our moral and psychological horizons, than to rail against it.

History: from John Hunter to John Peel

The way the procedure emerged and evolved has something to teach us. One might suppose that such a relatively simple practice would have been in use for thousands of years. It is just possible that it has. There is some evidence of its use amongst the Arabs of the fourteenth century, either in stealing semen from prize stallions or in wilfully impregnating rivals' mares with sperm from defective animals. Its origins, then, may lie in the deviousness of desert horse-traders — hardly an inspiring thought. More definitively, artificial insemination is usually identified as an 'invention' of the eighteenth century.[5] Professional medical opinion asserted that in fact conception could occur without sexual pleasure, a notion unthinkable in and to earlier generations. Nature, some doctors said, would accommodate a little artifice.[6]

In the early eighteenth century biologists attempted the insemination of frog, fish and insect eggs outside the body. That is, after all, where it normally occurs. That mammalian species also had eggs seemed likely, though they proved phenomenally difficult to find. The question remained in doubt until 1827 when the German biologist Carl von Baer reported his observations on a friend's bitch. However, in 1776 a Scottish doctor working in London, John Hunter, had already arranged artificial insemination for one of his clients. This only came to light six years after Hunter's death and twenty-three years after the event when his executor went through his medical papers. Hunter, a renowned anatomist and comparative zoologist, a medical teacher and surgeon, with a huge practice in London, had been consulted by a linen-draper because of anatomical abnormalities that affected his ejaculation.

> After the failure of several modes of treatment which were adopted, Mr Hunter adopted the following experiment. He advised that the husband should be prepared with a syringe fitted for that purpose, previously warmed, and that immediately after the emission had taken place, it should be taken up by the syringe, and injected into the vagina, while the female organs were still under the influence of coitus, and in the proper state for receiving semen. This experiment was actually made, and the wife proved with child.[7].

This then was AIH, not AID. Hunter gave it no publicity at the time, although quite possibly he and others were moved to help others in the same discreet way. Moreover, his role was simply advisory. The husband carried out the procedure himself, and, as client, retained control of the procedure. As with a number of other doctors over the next two hundred years, Hunter kept his actions to himself, the details only reaching his colleagues some years later.

In the early nineteenth century a French doctor, Girault, tried artificial insemination for a young Countess. This was done by Girault himself on 5 June 1838. It was, however, another thirty years before he felt able to publicise this and another eleven cases, of which ten had apparently been successful. Two years earlier, in 1866, the medical journals in Britain reverberated with the furore created by the publication of a book on sterility by the American doctor, J. Marion Sims, a surgical pioneer of terrifying callousness, then working in exile in the UK. Before the Civil War had forced him across the Atlantic, Sims had attempted artificial insemination with some of his patients. In France in the 1870s and 1880s several doctors made it clear that they too had followed this procedure, though it was 1887 before anyone explicitly referred to the possible use of donor semen.

Conservative medical opinion was nonetheless revolted by these reports, and one British journal had Girault's book reviewed in Latin, presumably to protect the image of medicine from sensationalist attention. Nonetheless, by the last decade of the nineteenth century artificial insemination had become an established practice in France, and in various ways had come to the attention of the general public. There were also doctors in Germany and Italy with some experience of it.

In 1908 a major controversy erupted in the United States over a case where, it was alleged, sperm donation had been used.[8] An American physician, Dr Addison D. Hard, claimed that as a medical student at Jefferson Medical College in Philadelphia, he had been involved in an attempt at 'artificial impregnation'. A wealthy merchant and his wife sought advice about their infertility from Hard's teacher, Professor William Pancoast. Investigation showed no reason for the woman to be infertile, but the man to be producing no sperm. Hard claimed that Pancoast discussed the matter with his students. The group proposed that they should elect the 'best-looking member of the class' to act as a sperm donor. This was done and the insemination performed under anaesthesia, without knowledge or consent of the woman or her husband. When the woman was found to have conceived, her husband was told what had actually happened. He was pleased, but asked that his wife should not be told. All this is supposed to have occurred around 1884. Pancoast died without revealing any details.

After twenty-five years, the sperm donor, and we must assume this was Hard himself, visited the merchant's son and told him the full story. Hard then published details of the case in a popular medical journal. Uproar ensued. It is an extraordinary story of male collusion. The actions of doctor and donor now seem rather difficult to justify. Hard claimed that some of the details were fictional. He had, he said, embellished some points, 'to set men thinking', although the essentials were true. It must have set women thinking too. Even if much of Hard's story is fable, his picture of what doctors in Philadelphia might be supposed to have done is depressing. It also suggests that sperm donors' later thoughts and actions may be rather complex and opaque to the men themselves.

It is difficult to determine whether and how often AID was performed in the first few decades of this century. There are papers in the medical journals that indicate that a very few doctors would do it occasionally, very discreetly, because of the continuing strong under-current of disapproval. In 1927 Schorohowa, who practised medicine

24

in post-revolutionary Tashkent, stated that she had found only 88 authentic cases in the world literature.[9] She claimed to have performed 50 inseminations herself from 1917 to 1925, having been consulted by 586 women with fertility problems. In only 3 cases was donated sperm used. In 22 cases children were born.

In Britain two of the pioneers were also women. One of them was Margaret Jackson, who worked as a doctor in Devon for some fifty years. In 1930 she started a clinic in Exeter to provide contraceptive advice. But by 1943 a third of the 500 women coming for advice believed themselves to be infertile. For some of them Dr Jackson arranged AID. The first child conceived after AID was born in 1942.[10]

Another woman in the UK who worked with AID from the 1940s was Dr Mary Barton, who as a doctor in India had seen the misery of infertile women. In 1945 she and her husband, Berthold Weisner, who was a biologist, and a medical colleague published an article in the *British Medical Journal* on their work, which produced considerable critical comment from doctors, from churches and in Parliament.[11] The donors were all married men, with at least two legitimate children, between 30 and 45, and some of them were used many times over.

The whole idea was not well received by the majority of their colleagues. Within four years a commission appointed by the Archbishop of Canterbury had recommended that the provision of AID be made a criminal offence. Nonetheless, since that time the practice has expanded slowly in several countries, despite continuing condemnations from various churches. In 1960 a committee in the UK under the chairmanship of the Earl of Feversham was set up to consider whether changes in the law were necessary in order to stop AID.[12] They decided against such a ban, although the committee did not go so far as to endorse AID. Requests for AID continued to be made to doctors through the 1960s. In 1971 the British Medical Association set up another committee, chaired by Sir John Peel. As the moral baton was passed from archbishop, to earl to knight, attitudes can be seen to have changed from strong disapproval to unenthusiastic acceptance, hedged around with concern for the legal position of the doctor. The publication of the British Medical Association report in 1973 marked the full entry of AID into the NHS system and one clinic was established the following year.[13]

Considering 'good practice'

Surveys now reveal a considerable degree of public understanding and support for AID, though not as much as for adoption.[14] Where once there were just a very few pioneers, now around 40 doctors in the UK are involved with AID. The figure in the US is at least 10 times that, for a population only 4 times larger. Professional and public discussion is rarely concerned now with the permissibility of the procedure as such but with how to handle the legal and psychological complexities. This is certainly the position taken by the most recent UK government committee to consider artificial insemination, the Warnock Committee, which reported in July 1984.[15] The phrase in their report which stands out in the discussion of AID is 'good practice', the emphasis being very much on imposing and maintaining professional standards.

Let us consider what doctors say they do. In 1977 a nationwide survey was conducted in the United States of physicians thought likely to be practising AID.[16] 379 said they did so. Probably as many again actually do so, but were not contacted in this survey. The majority of this group (264 doctors) handled only a few cases a year, and only 5 were involved with as many as 50 to 60 or more cases a year. Whilst they were to be found in 46 American states, the major concentrations were in New York and California. The situation in Britain shows a similar geographical imbalance. This means that some people have to travel long distances to get this kind of help.

Donors were almost always chosen by the doctor. Of doctors 92 per cent would not allow people coming to them to choose, offering them instead sperm from an individual with a few specified characteristics. Overall 62 per cent of doctors selected their donors from local medical students; 10 per cent used other university students. Overwhelmingly, then, donors would be in professional training. Beyond this there was little further screening. Although most doctors said that they would reject donors on certain genetic grounds, few of them actually took steps to discover whether what donors told them was true. Most were paid between $25 amd $35 for each ejaculation. When asked whether there was a limit on the number of pregnancies that would be established using the sperm of any one donor, a significant number of doctors in the survey simply ignored the question, suggesting that in fact they had no such limit. But of those that replied (249 out of 379) three-quarters said that there was a limit of 6 pregnancies. One doctor said that one donor had been used in 50 pregnancies.

Whereas most doctors followed the course of the pregnancy, even if they were no longer professionally involved with it, only a third kept records of the donors. Overwhelmingly this was because of a concern to protect the privacy of the donor. Alternatively some mixed sperm from several donors for each insemination or used them alternately in each cycle, to make it harder to identify the particular donor who was the genetic parent. That would make it harder, if not impossible, to pinpoint deficiencies in technique or carriers of deleterious genes passed on to the AID child. Although 95 per cent of the inseminations were done because of a husband's infertility, at least a third of the doctors had performed AID because of a risk of genetic disease, and 10 per cent had done so at least once so that a single woman could have a child.

This survey revealed several worrying features. For example, there is the question of how many times a donor is used. Getting volunteers is not easy and results vary between different men, so there is a temptation to stick with a few particularly fecund donors. The problem is not that donors get worn out or become depraved and corrupted but that using the same donor many times increases the chances of accidental consanguineous marriages. This would arise if women are unknowingly given sperm from a relative or if people in the next generation marry someone who is in fact their half-brother or sister. Now the risks of this happening are often overstated. It is very much more likely that it would occur through unacknowledged paternity derived from extramarital sex. For example two women may, unknown to them, have had the same male lover and have had children, who were always assumed or said to be the children of another relationship. Nonetheless in small communities the statistical risks of too great a reliance on a given donor could be significant. For this reason the Warnock committee recommended a limit of 10 pregnancies per donor, to be enforced by a Licensing Authority. How they would check this is not clear.

Then there is the question of genetic disease. Such conditions arise because of the inheritance of traits from the preceding generation. You do not contract a genetic disease by infection, you develop it because of your genetic make-up, which derives from contributions from both your genetic parents. Genetic diseases are basically of two kinds, those called recessive conditions which only arise if you have inherited the trait involved from both parents, who are called 'carriers' of the trait. Almost always they will be unaffected themselves by it in any significant way. Then there are so-called dominant conditions, for

27

which you need only inherit the problematic hereditary trait from one parent. It follows of course the affected parent will himself or herself suffer from the disease, though they may not have done so when they conceived their children or known that they would do so. Finally, genetic diseases are usually very serious, and kill those suffering from them at a young age. But not always; one dominant condition, Huntington's Chorea, only manifests itself in middle age and a few genetic diseases can now be controlled for long enough for some people with them to reach reproductive age.

How does this relate to AID? For some genetic diseases unaffected carriers can be identified. This we shall discuss in Chapters 8 and 9. If two carriers marry and wish to have children, they face a risk of producing an affected child with each pregnancy. In the case of recessive diseases this would be 25 per cent. For some diseases diagnosis in pregnancy is possible, so that affected foetuses may be aborted. But an alternative for those that find this option unacceptable morally or psychologically is to arrange conception by AID, using sperm from a man who is known not to be a carrier of the trait concerned. Or it may be that a man knows that he suffers from a dominant genetic disease. Almost certainly his wife will not suffer from the same disease but if they have children together there is a 50 per cent chance that the child will also suffer from the disease. An option in this case would be AID. As we shall see later on, more and more people will find themselves in this situation in future, although that is not to say that they will necessarily have recourse to AID. Whether they do so will depend on their attitude to abortion, to the kind of suffering that genetic diseases create, and to AID. Clearly AID only makes sense in this context if one knows that the donor is not a carrier of the particular trait concerned. That is the point of finding out whether doctors actually check the genetic history of donors.

The more we learn about human genetics, and the easier it becomes to identify carriers of particular genes, the more necessary it becomes to know the genetic status of sperm donors. The same will also be true with ovum and embryo donation, which we consider in Chapter 4. The vast majority of us have no idea whether we carry a few deleterious recessive genes. But as time goes by, more and more people will know and that will affect the kind of donor they are prepared to accept. At the moment ignorance is bliss. Similarly, as it becomes technically easier and cheaper to make such checks on both donors and recipients, doctors' moral and legal obligations to do them will strengthen.

28

Such 'quality control' is not trivial technically; nor is it immensely difficult. It requires a modern laboratory. This may mean linking AID even more closely to the hospital system. That means placing it more under the control of the medical profession, which certainly creates a problem if one wishes to maximise access to AID.

One possibility would be to follow the French approach, which is centralised and regulated.[17] In the 1950s and 1960s in France AID was done privately and very discreetly by physicians in private practice, because of strong disapproval from the Catholic Church. In 1973 two sperm banks were opened in Paris, marking the acceptance of AID into the public health system. These were the first units in a nationally co-ordinated network of such centres called CECOS (Centre d'Etudes et de Conservation de Sperme), which by 1983 had expanded to include 19 centres. The ten years since 1973 saw a considerable increase in the number of requests for assistance, from 278 in 1973 to 3,021 in 1982.[18]

Donors are not paid, because it is thought that payment has mercenary connotations which discourages participation. Instead donors are encouraged to see their donation as an act that affirms their membership of society at large, a gift that acknowledges the needs of some in the community. Furthermore, the gift is represented as being from a 'donor couple' to a 'recipient couple' to minimise any connotations of adultery. The implied link is between families and not between individual partners in separate couples. This high-minded campaign of moral re-education is backed by considerable research to improve the success rate of fertilisation. Both donors and recipients must be married and are assessed medically, genetically and psycho-logically. The extent of control also over local standards and policies is deliberately firm: protocols define how insemination may be carried out in each centre and detailed records of donors and recipients are kept on a national computer system.

The organisers of this network have as their goal the creation of a new social understanding of the biological facts of AID. They want to imbue them with a different significance, by an intervention in French culture to make AID not merely something to be tolerated but almost a social duty and a sign of enlightened attitudes. People are encouraged to see AID as principled, responsible gift-giving, whereas in other cultures donation is marginalised as the inconsequential action of students with no commitments and no kin. CECOS deliberately uses the symbolism of family life, procreation and marriage to get across the message that participation in the scheme is not demeaning or polluting, but something to be valued by donors and recipients. At the same time it represents a pervasive system of moral control over who has access to AID.

Another approach to AID, that in some respects at least is very different from CECOS, has been pioneered by Lesbian women in Britain, America, Holland and elsewhere.[19] They reject the idea of casual heterosexual intercourse to conceive, but face almost universal criticism and obstruction if they seek AID. Recently a few women have decided to take matters into their own hands, because they see the issue not as one of finding a doctor with different values, but of creating the conditions within which women can act autonomously. They also reject the connotations of 'artificial inseminaton', preferring instead to emphasise women's control of their own bodies, with the term 'self-insemination'.

The donors are often gay men, who share the view that parenting need not be restricted to heterosexual couples and who also wish to pursue a life-style based on a minority sexual orientation openly and freely. The symbolism of affirmative gift-giving from men to women parallels that in the CECOS arrangements, despite otherwise deep differences in ideology. Self-insemination groups deliberately and forcefully reject the ideas of family, marriage, professional expertise and expert control of reproduction so evident in the French scheme.

'Good practice' is inevitably going to mean different things to different people. For some it is an unproblematic ideal of professional behaviour to be administered with great respect by fellow professionals, in order to outlaw the cowboys. For others the whole idea that professional men should try to exercise this kind of self-regulation, selection of clients and moral control is anathema. There is a real danger in assuming that the gingering up of technical standards is enough. It would be all to easy to put artificial insemination back in the professional closet, because it is too challenging to consider more openly. In a sense it is a threat to traditional notions and habits and some people will react accordingly. However, I prefer to call it an opportunity to think about the difficult issues of sexual orientation, the obligations of parenthood, children's rights and needs and reproductive autonomy. All the same issues will come up in one way or another with the fancier technologies of in vitro fertilisation, sex predetermination, and antenatal diagnosis. The crucial point about AID is that it is so much simpler technically. It cannot be kept under the control of professionals, even if we thought it for the best. It is far better to try to face up to some of the undeniable anxieties that it

30

creates, rather than try to repress them. This thought underlies my approach in the rest of the chapter to the specific issues of self-insemination, disclosure to children and the banking of sperm.

Self-insemination considered

This has two virtues, which deserve consideration, before self-insemination is dismissed out of hand. The first is that it builds people's confidence in their understanding of their bodies and in their ability to take informed medical decisions. Women who have been involved with it testify that it leads to an increased sense of responsibility for their health and well-being. It reverses some of the alienation that modern medicine commonly induces, making people feel more involved with their own health care. It therefore embodies ideals that many support. The second virtue is that it forces us to think about who is allowed to be a parent. Why should AID only be offered to married, infertile, heterosexual couples? If nothing else self-insemination is usefully provocative.

Self-insemination by single or Lesbian women removes intercourse with a man as a precondition for parenting. Undoubtedly that makes some people very uncomfortable as some women slip around the constraints that biology was thought to impose and find their way into parenthood. The common objections to this concern the material difficulties of being a single-parent, the effects of the absence of a male father-figure from the household, and effects of the mother's relationships on the child's own sexual development and social interactions. The first applies only to mothers on their own, the second to both cases, and the third can be restricted for the sake of simplicity to Lesbian mothers.

It is very tempting to say at this point that all kinds of families exist, that divorce, family violence, separation and single-parenthood after marriage are not uncommon, and that married partners maintain all kinds of relationships under a cloak of conventionality, so why worry about a little more variation? Tempting this may be, but it is a silly and specious argument. The fact that many things happen does not mean that any are necessarily good. If we have a choice we should try to do better. Or one might say that the question of who can be a parent does not arise for people conceiving without AID, so why raise difficulties for people who are forced into it? This is much more persuasive, but

31

still not acceptable. If anything it is an argument for more forethought all round.

As far as women living on their own are concerned, the acceptability of AID is secondary to the question of seeking to have children on one's own. If bringing children into the world like this is defensible, surely not seeking a casual sexual encounter to that end is a sign of self-respect? But what of single-parenthood? If women have reasonable financial resources, access to some kind of help with child-care, and experience of significant, adult relationships, then we must say that they can handle the challenge, if that is how they choose to live. AID should be made available to them.[20] But some people will still say that the interests of the child are not well served by the absence of a father. That may be true, but the real question is what the woman plans for the future. If she has in mind truly minimal social interaction with men, then there may well be problems. On the other hand, if a married couple planned a pregnancy knowing that the man was going to be absent a great deal on business, I doubt if we would criticise them.

The most obvious conclusion is that women taking this step need some advice from friends and people who can counsel them without threatening them. One of the virtues of self-insemination is that, unlike casual sex, it is unlikely to be obtainable without some discussion. The sub-culture which sustains it simply does not work in an offhand way. In that sense self-insemination is a misnomer, since it requires the cooperation of others. In any case adoption by single women is possible under British law, and the legislation was drafted with single-parent adoption in mind.

The case of Lesbian couples is both similar and different. The would-be mother is not on her own: she is likely to be in a stable relationship, possibly living with her lover. So the economic and child-care problems should be less. But what about the absence of a father and the complication of minority sexual preferences? There are in fact some studies of Lesbian couples and their children which offer some guidance.[21]

They differ in method and approach, and, to some extent, in their findings. Some studies asked whether children seemed to 'cope'. Others tried to assess whether the children developed 'normally'. Despite this rather pessimistic start, the researchers report some very positive findings, as well as some problems. Some children clearly understood and respected what their mothers had done, even though they themselves had or expected to have heterosexual relationships. Some were resentful and confused, but that was less common. Almost entirely the mothers were women whose marriages had failed, after having had children, and who then formed an intimate relationship with another woman, who joined the household. The children's experience of their mother's changing sexual orientation was thus powerfully conditioned by divorce and strife at home, followed by the arrival of another adult. But what stands out from the interviews is that the mother's sexual orientation was very much a secondary issue, compared with the experience of the divorce and the reconstruction of a loving home environment. Moreover, the women were far from unconcerned about the absence of a father-figure and often took considerable pains to see that their children maintained close contact with their father or with other adult male friends.

These studies provide valuable evidence that women in Lesbian relationships should have access to AID, with some discussion and counselling. They should be able to enjoy this opportunity without the fear of losing their children which often haunts them at present. There is also room for more research. Rather than asking whether the children end up with conventional sexual preferences, as a sign of being 'normal', we should inquire whether they seem able to understand and deal with unconventional sexuality. After all, it would be a very good thing to transcend 'normal' people's limitations in this respect.

The psychological and developmental case against self-insemination for Lesbian women looks pretty thin, when you substitute factual evidence for homophobic stereotypes. Minority sexual orientation should not disqualify women from seeking to be parents. Now let us go back to the other virtue of self-help: taking responsibility for one's medical care. In the case of self-insemination, the question is really how some kind of technical input is to be arranged. To be completely informal and never assess how self-insemination is being done has real problems.

The most troubling problem at the moment is the virus that causes AIDS, which can be transmitted in semen. A paper in *The Lancet* in September 1985 reported that four women in Australia had been exposed to the AIDS virus, through having received donor semen from a clinic.[22] One of them had symptoms which may develop into AIDS and the other three had antibodies to the virus in their bloodstream. Three of them had become pregnant, but neither their children nor their husbands showed any sign of having been exposed to the virus. Most donors are now screened to see if they have antibodies. They are re-tested after six months, and only if that test is negative too is their

33

semen used. This will restrict the supply of donors and prevent the use of fresh semen. But it also means that medical data can be used to select donors, rather than inferences from life-style.

It makes sense then to use semen from tested donors. This is probably beyond the resources of the informal network of gay men and women, who have tried to help each other. One possibility would be for women to create their own clinic, with laboratory and storage facilities, backed up by a women's counselling network. The Warnock Committee proposed that a Licensing Authority should be set up to monitor the practice of artificial insemination. This is politically and technically desirable as long as such a body has the vision to appreciate the value of a women's self-insemination clinic, which could provide non-alienating technical advice.

Secrecy

If two women live together, or one woman lives independently, at some point their children are likely to ask how they were conceived and where their father is, if they have not been told. But a man and woman who have used AID can misrepresent the details of paternity if they so choose. As we have seen, many do. The same must occur with extramarital affairs which are a traditional alternative to AID.[23]

Since there is nothing for the child to remember, unlike the experience of being adopted, one might say that the details of paternity are best misrepresented. But the interesting thing is that the psychology of forgetting and denial is not that straightforward, and keeping a secret over many years becomes a strain, which tells, literally. In moments of anger, things get said, or the fear that the child himself or herself has guessed can become a real problem. Or children may just sense that there is something anomalous or mysterious in their past, without having any idea what it is. So some people have come to argue for evolution towards greater openness, as parents are encouraged and helped to talk to their children about these things. But at a number of levels it is not that easy.

Two analogies come to mind. One is adoption, where the trend is very much towards disclosure. 'Good practice' nowadays in counselling future adoptive parents emphasises telling children that they were adopted, how and when this occurred, and what it means. The intention is to re-affirm bonds of love and affection, after the disclosure that 'other parents' exist, or existed. Making this revelation to a child will mean that he or she will certainly think about and sometimes go to find his or her original parents. This prospect is worrying for all concerned, but the evidence is that such openness works, and that no great disruption occurs. Only a small proportion of people who were adopted as children actually try to find their original parents.[24]

How far can we pursue the analogy for AID? Would people be enabled or encouraged to find the man who acted as a sperm donor? His motives and feelings would have been very different from those of a woman or couple giving up a baby for adoption. Women who make their children available for adoption are not baby donors, in the sense that we speak of sperm donors. But, for the person conceived after AID, it may well be helpful to know something of or perhaps meet the male genetic parent.

My other analogy is the practice of telling children that they resulted from an accidental pregnancy or a decision not to abort. Here there are no genetic complications to negotiate, but what is revealed, in anger or lightheartedness, is the absence of a desire for a child. Adopted or AID children are desperately wanted, whereas in this case desire and anticipation were not there. Complete honesty about the circumstances of conception may not be the best policy. Not everyone can present their honesty in a loving way.

The maintenance of secrecy after AID is not a good idea, but people need help to be more open. Many people seeking AID do not share this view and are very resistant to the idea of counselling, to help them come to terms with their feelings. Elizabeth Alder's study of such couples in Edinburgh in the early 1980s revealed that most of them saw any kind of discussion as a test of 'fitness for parenthood', which they resented.[25] Often they had experienced this when applying to become adoptive parents. The men made this particularly clear, their feelings of masculinity having been damaged by the discovery of their infertility, and the last thing they thought they wanted was to discuss them with a social worker or counsellor. On the other hand, many of the people interviewed by Snowden and Mitchell, in most cases some years after having had recourse to AID, said, 'If only there had been someone to talk to'. For them, secrecy had become a torment, even though initially it had been so eagerly sought.[26].

There is another problem with disclosure, with both practical and psychological ramifications. If you tell someone that he or she was conceived with donor sperm, should you also tell them something about the donor, and even who he was and where he is now? This would be practicable now in only a very few cases, where the records exist. We could make it 'good practice' to keep such records, where at the moment it is thought good practice to destroy them. If we decide to make this change, then we change the nature of donation, and its meaning to donors. It is often said that if their identities were to be revealed, then men would become extremely reluctant to volunteer, for fear of possible, future psychological, social and legal complications, involving not only themselves but their families. In Sweden, new legislation requires such data to be kept, and it will be interesting to see what actually happens there. Until now the evidence about what donors feel is somewhat mixed, but little data is available. One study of 67 donors in Melbourne showed that half would not participate if their names were available to parents.[27] At the same time only 15 per cent thought that their names should actually be available, which is a slightly different question. When pressed on this point, 18 per cent said that they would not object to their names being released by mutual consent. 15 per cent said that they were reticent about the idea because of possible legal complications for themselves. Arguably this feeling could change if the law were to be altered. The other interesting issue to emerge from this study concerns donors' attitudes to the children their donations make possible. Despite what they are expected, even required, to feel by the people running this AID programme, many of them were curious about 'their' children. One can interpret these results in various ways. Some men expected complete anonymity. But nonetheless keeping more detailed biographical information would not be problematic for most; and some donors were prepared to meet 'their' offspring, when they had reached the age of 18. This is in sharp contrast with the overwhelming disinclination to meet the child's parents.

As the medical utilisation of genetic data increases, it is going to become more and more important to compile and record such data about donors. This information will surely be revealed to donors themselves, as will an indication as to whether they are fertile. As we have seen, they are also going to have to be tested for AIDS. The trend is clearly towards donation having many more medical implications for the donor. Rather more is going to be asked of him than a little masturbation at lunchtime.

I would go further and say that donors must accept that some biographical information will be compiled about them and made available to people born after AID at age 18. If donors cannot handle that prospect, they should not be asked to participate. That has two further implications. One is that donors need to be approached more carefully and given the opportunity to discuss what they are doing. This takes their role somewhat more seriously, treating them in a less furtive and embarrassed way and acknowledging that they will have feelings about their actions, which can usefully be explored. Having said that, I still think that donors should remain very peripheral figures and should be selected for their altruism by not being paid.

The other implication concerns the legal responsibilities of donors and the law of legitimacy. At present people born after AID are technically illegitimate under English law, because their genetic parents are not married to one another. In practice parents often enter the name of the mother's husband on the birth certificate as the father in an act of perjury. In recent years the whole notion of 'illegitimacy' has come under critical scrutiny.[28] It is argued that the distinction has become entirely unnecessary, although it remains a source of stigma and embarrassment. This is particularly so as more and more people are having children without feeling the need to marry. The residual implication that the child was the result of an illicit liaison is thus entirely out of place. In 1982 the Law Commission in the UK recommended that the category simply be abandoned, so that whether or not a couple were married would not count in characterising their relation to the child.[29]

However, this immediately creates problems for AID, unless one can find some way of indicating that the donor cannot be considered to have paternal responsibilities. It is desirable, for example, to distinguish a birth after AID from one occurring after adultery, where the biological father might be held to have some legal obligations. The Law Commission concluded that a special case should be made for AID children and recommended that the parents should be defined as the biological mother and her husband or partner, unless he could prove that he had not given his consent to the artificial insemination. This would then resemble the situation in France, Netherlands, Portugal, Switzerland and many US states. These recommendations were endorsed two years later by the Warnock committee.

Sperm banking

Finally we must consider the practice of sperm banking. For a few thousand pounds one can set up a refrigerated sperm bank – in the US they are often run as private businesses, linked into chains across the nation.[30] Perhaps someone is even selling franchises? You can make a deposit on the way to work, and by the time you have your morning coffee your sperm will have been washed, checked for motility and contamination, split into lots, decanted into special freezing straws and placed in the freezer. Apparently the Vietnam war boosted the demand.

Often this is really another form of AIH. John Hunter advised his patient on how to circumvent his disability. In the twentieth century people seek a solution to possible future infertility. If this was all there was to it then it would be sufficient that sperm banks should be run efficiently, with records kept in order, samples not mixed up and confidentiality maintained. But suppose that the man who makes the deposit dies and his wife or lover then asks to be inseminated, what is the right course of action?

This problem has come up recently in France where a Mme Parpalaix eventually took CECOS to court to get them to release her dead husband's sperm. In the autumn of 1985 there was a case in Manchester where an NHS clinic received a similar request. In France CECOS were ordered to release the sperm they held, though the court did not order that Mme Parpalaix had to be offered artificial insemination.[31] I hope that in these circumstances, despite the litigation, the CECOS doctors were prepared to help her, or at least to hand over the sperm at monthly intervals, for her own use.

But some people would certainly say that such desires should not be met, because this is the wrong way to come to terms with death and is not in the best interests of any future children. Obviously this was what some of the French doctors felt. If you take that view then the question of what the deceased man may have wanted is irrelevant. You could go further and order sperm banks to tell potential depositors that this policy would be enforced. But is this view correct?

There are several questions to be examined here. Firstly, there is the absence of the father. In effect this is another, but special, case of deliberate single-parenthood. Children are brought up now in single-parent families, where the father has died very soon after conception, and before the birth of the child. However, with insemination the woman is deliberately trying to conceive after the death of her erstwhile partner. The possible economic or social difficulties are not obviously severe enough for us to say that she should not be allowed to go through with this. Any argument must be largely psychological.

Secondly, there is the psychology of grief and mourning. In the weeks and months after the death of a husband or lover thoughts of re-creating them may come to mind. Some people would, quite reasonably, argue that this desire should be resisted, in order to become able to form new relationships and transcend the past. Women contemplating this step should be given the opportunity to discuss it at length. But respect for their autonomy demands that ultimately the decision is theirs. The obvious parallel is with a woman in the emotional turmoil after someone's death seeking to have a child through casual sex. Most people would say that this was misguided. Good friends would counsel caution, or contraception, but it would not be something that one could actually prevent. Nor would anyone withdraw medical care from her, either prior to conception or during pregnancy, though no doubt various people would try to hand out some punishment. So in what sense is the situation of the woman seeking insemination different? One difference is in the procedure. Self-insemination is a sign of greater self-respect – and also of greater respect for others – than sex, which is seen basically as a means to covert procreation. Another difference is what one would tell the child. Surely it would be better to say that both parents had planned the child's conception before the man's death, rather than have to admit that in the aftermath of one person's death an irrelevant stranger had been picked up in a disco for a one-night stand?

In the US there is at least one sperm bank of a different kind. This collects and makes available the sperm of men selected for their exceptional academic credentials, including in a few cases, the award of a Nobel prize. This project is hereditarian; that is, its founder believes that intelligence is an intrinsic capacity that is inherited to a large extent and that a woman inseminated with the sperm of a man of high academic achievement will have a brighter than average child. It is also eugenic, in that the people involved with the Repository for Germinal Choice believe that this selectivity is a good thing, because it could increase national or racial levels of intelligence. In his book *Out of the Night*, Herman Muller argued that responsible citizens would in future practise 'voluntary germinal choice' and use the sperm of outstanding individuals when they wished to have children. Before he died in 1967 Muller met the Californian millionaire, Robert Graham, who ran a company making plastic spectacle lenses. In 1971 Graham created the Repository for Germinal Choice and later set up the Foundation for the Advancement of Man [sic], which now finances the Repository.[32] Graham and an assistant solicit donations from selected individuals. For example, in 1981 they approached all the Nobel prizewinners from that year and were turned down by all. But sometimes, at a less exalted academic level, they strike lucky. They then visit the man to collect his donation. An extended list of academic and medical details is compiled and a photograph taken, although each donor remains anonymous. These details are made available to women coming to the Repository. They must pass a test to show that their IQ is at least 130, and they may choose a donor for themselves. They may also switch donors between inseminations. Several children have now been born from this project.

This little exercise is very depressing, not least because of the very shaky theoretical foundations on which it rests and the apparent myopia towards the ways in which exaggerated parental expectations can create anxiety in children. The nature, development and variability of human intelligence is so little understood, that it is much better not to go to these lengths in the hope of marginally improving the attainments of one's children.

Interest in this sperm bank continues despite withering criticism from leading scientists. François Jacob, Nobel laureate in 1965, remarked that only someone who had never met Nobel prizewinners would consider trying to create lots of them. Clearly, this exercise panders to the vanity of highly privileged men, who ought to know better. I would also be a little concerned about the motivation of women selecting this option and anxious about their relations with future children. But autonomy is autonomy. If we consider that self-insemination is valid, then some people will use their freedom for misguided eugenic ends. We do not have to keep silent about the lack of sense to their actions, but we cannot stand in their way.

AID is likely to become more common in the years to come. For two hundred years, if not for longer, it has been shrouded in secrecy, as a private piece of magic enacted by a doctor. That kind of evasiveness, which makes the continuation of incompetence possible, is not consistent with expansion or an enlightened psychology. We now have the opportunity to take the 'facts' of paternity in such cases, and, without denying or repressing the genetic relations they imply, set them in a new context where the social relations of parenthood are acknowledged to have more power and more substance. In that way we could come 'out of the night', but not into Muller's eugenic dawn. We would not be opting for something that goes 'against nature', but for something profoundly human in our conscious choice of what significance is to be conferred upon natural events.

The only sensible and mature response to the imminence of morally problematic procedures and technologies is to anticipate their consequences, both good and bad, to find analogies with things we already understand. This will allow us to develop the confidence to handle the choices they offer and to reject those which are profoundly offensive. This confidence is something that must be built from the bottom up. That is why Mr Graham's little repository is such an insidious idea, because it plays on the anxiety of people that they will be unable to raise intelligent and successful children without help from his bank.

Chapter 3

In vitro fertilisation:
experience and technique

AID is in essence so simple, although we now think it necessary to add several complicated checks to the basic procedure. It is ironic that we designate it artificial, when mechanical intervention forms such a small part of the whole process. With in vitro fertilisation there is a quantum leap upwards in technical complexity. Bringing about fertilisation outside the human body is much more ambitious. It took far longer to achieve and the technique admits of a whole series of variations, with their own particular moral implications. Originally it was intended to circumvent a major form of female infertility. This remains its principal purpose, but the variations·already embrace certain forms of male infertility and will soon be applied to the avoidance of genetic disease. Its potential scope, then, is enormous.

Ivf is no recent arrival on the front page or the mid-evening news. The technique is established, and many of the issues that it raises could have been tacked at least twenty years ago. There is therefore an air of contrivance about recent controversy. In particular the question of what kind of research, if any, involving human embryos is permissible has attracted a great deal of public discussion very recently. In Britain there is now palpable concern with how embryos are used. I shall make a case for such research, regardless of whether we think of ivf as worthwhile. Embryo research does not need ivf to justify it.

In this chapter I am concerned with exposition. I discuss the biology of fertilisation and I describe ivf as a procedure. In the next chapter I describe briefly how ivf developed. A long period of research, that was socially almost invisible, led to the more public activity of a few pioneers, and on to the present intense professional interest. In the process very few of the moral and social issues were sorted out, so that there is now something of a scramble to do so, in the face of proliferating technical possibilities. In Chapter 5 I discuss some of these problems and offer my own evaluation.

The biology of fertilisation

Human beings, as we learn at school, in the playground if not in the classroom, are created through the fertilisation of an egg by a sperm. The largest cell in the human body is entered by and fuses with the smallest cell. Two sets of genetic information are mixed together and a cell with new properties and potential created. When it divides, we call it an embryo, and to many people it seems straightforward to regard this dividing cell cluster as a new individual or human being. In fact the biological situation is not so clearcut. It is not at all obvious where to mark a beginning or where to place the origin of individuality. Fertilisation is not a discrete event, but a process. It is extended in time, and its outcome is somewhat variable. [1]

It is not like switching on a light, which one moment is off, and the next on. It is more like switching on a computer, where low-level systems activate higher-level systems, and programmes are called up from memory, so that over a finite period, a fully functional system is in operation. But even this metaphor scarcely captures the complexity, balance and beauty of early embryological development. In the human case things go much more slowly. Fertilisation takes several hours and is an ordered sequence of events.

Eggs are made in the ovary. The first form within the embryo, during pregnancy. In a sense then every pregnant woman represents three generations together. These immature eggs, or oocytes, remain in a latent state for years. After puberty one or two mature in each menstrual cycle and become ready for fertilisation. At ovulation the ripe egg is released from one of the ovaries and enters the Fallopian tube. This is where fertilisation normally takes place.

The mature egg is surrounded by cells from the ovary, and a ring of non-cellular material, called the zona pellucida, and an exterior membrane. In the Fallopian tube sperm undergo two changes, one called capacitation, the other, the acrosome reaction, which releases enzymes that assist entry to the egg. Sperm first bind to the zona pellucida, and then one or two try to move through it and on through the outer membrane. All this takes time. When one reaches the interior or plasma membrane this sets off changes that prevent more than one sperm getting through. Once inside the egg the sperm head expands and forms what is called a pronucleus.

At this stage, unlike the sperm, the egg still has a complete set of chromosomes. Soon after the eggs were formed these chromosomes began to separate into two paired sub-sets. But this always stops halfway. Since sexual reproduction is all about taking one sub-set of hereditary characteristics from one parent, and one from the other, one sub-set still within the egg is jettisoned. So one set of maternal chromosomes is expelled from the egg, and the other forms into another pronucleus. The two pronuclei in the egg then fuse, and the two sets of chromosomes within them form pairs. Once this has happened it is conventional to say that the process of fertilisation is complete. The next stage arrives when the fertilised egg – known technically as a zygote – divides to form a two-cell embryo.

The fertilised egg soon begins to divide and the number of cells doubles from 1 to 2, to 4, to 8; its overall size remains about the same. At this point, about 3 days after fertilisation, it enters the womb. As cell division continues its internal organisation starts to change, so that the embryo becomes in effect a ball of cells with fluid in the centre. It is then known as a blastocyst, and after further growth, is ready to become attached to the lining of the womb. This process of attachment is known as implantation, because some of the cells actually grow into the surface of the uterus. Others begin to form into a membrane that will surround the embryo and others constitute the embryo itself.

Not all blastocysts implant. In human beings attachment begins about 6 days after the initial events of fertilisation and the whole process takes about a week. One may define the end of implantation as the beginning of pregnancy, although chemical signals involved in the process may be detectable whilst it is going on. By this point the embryo has perhaps only doubled in size and is now about 0.2 mm across, so small indeed as to be virtually invisible to the naked eye. Nor, since it is just a ball of cells, is there even the vaguest resemblance to the human form.

This whole process cannot occur if the egg or the sperm cannot pass along the Fallopian tube. If the tubes have been damaged by infection, either after abdominal surgery or as bacteria have found their way upwards from the cervix, then they may be blocked. In this case the woman will be infertile. The eggs may well ripen with each cycle but can never be fertilised and will simply degenerate. Fertilisation and subsequent development also may not succeed if the ovary fails to secrete enough of the required hormones. Similarly, if too few sperm are produced, or they are insufficiently active to reach the egg, fertilisation will not take place.

This blockage of the tubes is not uncommon. The tubes may have been removed surgically, because of some earlier medical problem, or they may have been clipped or cauterised as a form of sterilisation, although the woman now wishes to have another child. One way to diagnose a blockage is to fill the Fallopian tubes with a dye that shows up with X-rays. If the fluid does not spill out into the abdominal cavity – the ovary is not physically connected to the tube, and hence should have an open end – then the tubes are blocked. This sounds simple, but it can be painful and awkward in practice. It is possible in some cases nowadays to reconstruct the Fallopian tube, by shortening or rebuilding it. This is done by microsurgery, with some success. If the tube is no longer there, reconstruction is not possible. Joining the ovary directly to the uterus, transplanting Fallopian tubes and using nylon substitutes have not worked.

Ivf as procedure and experience

Infertility is a technical term, as well as one in everyday use. The two meanings overlap. It is defined statistically in terms of an inability to become pregnant in a given period of unprotected intercourse, usually one year. The possibility of pregnancy is not excluded, as it is in sterility, but its likelihood is reduced, and one may or may not regret the fact. Intrinsically it is a state of uncertainty. This may be compounded by not knowing the reasons for it. Often people realise that it is taking them longer to become pregnant than they think normal or acceptable, and they seek medical help. That in itself may be traumatic, embarrassing and isolating. It may lead to a solution. But if the investigations and therapy continue they are likely to induce new stresses, both for the individuals seeking help, and correspondingly for their families, friends and doctors. [2]

Those who have gone through this experience speak of initial shock and denial, leading to deep and often recurrent depression, and a manifold sense of loss. [3] Men and women are affronted differently by these problems and in coping tend to make different demands of each other. Men typically withdraw and say little; women are more likely to require discussion from an increasingly reluctant partner. Combined with this may come a sense of isolation, if they know no one who seems to be having the least difficulty in conceiving, and alienation from and resentment of friends or strangers who have children. Even friends who try to be sympathetic can come to be seen as a threat and are abandoned. Nature itself seems alien and unpredictable, since if something as 'natural' as conception has gone awry, who knows what else may not go wrong?

These troubles are punctuated by the cycles of the woman's body, as each new menstruation brings new despair, and medical investigations bring intercourse to a related schedule, followed by periodic reporting of its frequency and timing. Very many subtle antagonisms, fears, fantasies and insecurities manifest themselves and have to be dealt with in some way. One important way of doing that is discussion, with each other, with others in the same situation, and with the most thoughtful friends. In recent years self-help groups have been formed in several countries to make this possible. Not for nothing is one American organisation called with admirable ambiguity, Resolve. [4] Nonetheless, people consistently report that they would like far more opportunity to talk about their feelings with their doctors. The medical response to this is generally, 'Well, we try, but we can only take on so much'.

This is the context in which ivf takes place, and any moral or social evaluation of the technology must take account of the hurt and anger and the commitment to go yet further into the medical maze that infertility brings. In several senses people qualify for entry to ivf programmes after years of medical consultations. Ivf is a symbol of hope and a new source of stress; a breakthrough that may make pregnancy possible and an experience with its own particular dangers.

Hope springs eternal at new technology, and ivf is no exception. Some see that as a problem because people form unrealistic expectations of the chance of becoming pregnant. They will, it is argued, fall prey to the hype that medical careerism and competition creates (see Chapter 4). Others maintain that the drama of ivf brings people to a useful resolution of their feelings, since the chance offered them seems so special that they know there is nothing more that could be done. This is an interesting argument, but it comes from people running ivf programs, who inevitably will want to feel that the people they are trying to help can cope. [5] The people whose feelings will not be studied will be those for whom ivf has failed. [6] As we shall see, there will be at least as many of them as those who will have children.

The first point of stress is acceptance into an ivf programme, which involves an interview, often after referral from another doctor. It is almost certain, even if successful, to mean a wait of several years, during which time some people withdraw. The selection takes considerable account of the medical history of both partners, but it is also clear that to some degree people seeking ivf have to 'qualify' psychologically, and be thought emotionally stable, to be able to handle failure, and have what seems a good relationship with their partner. [7] The scope for all kinds of prejudice and judgment based on stereotype here is considerable. Some ivf groups make it clear that they are somewhat selective on such psychological grounds, and feel they have to be. Others claim exactly the opposite. For those seeking ivf on a private basis, which is probably the majority in most countries at the moment, couples also select themselves economically. If they can only afford one cycle of 'treatment', this will in turn increase the psychological stress. Those who do not fit the sexual, marital and social norms will experience other forms of anxiety and resentment. Some programmes simply exclude them from the start.

A major part of the procedure involves determining the woman's ovulatory cycle, so as to know when the release of an egg is about to occur. [8] If the cycle is very irregular, or indistinct, then an order may be imposed by giving large doses of hormones, which sometimes have unpleasant side-effects. In most cases women are given a fertility drug to stimulate the ovaries to produce multiple ripe follicles. The work of the ivf team is prescheduled two months ahead and those who are to be dealt with in this time are told to contact the hospital when their next menstrual cycle begins. The injections of the fertility drug begin on day three and are continued daily. The couple is told to stop sexual relations after day five and from day eight the woman attends the clinic for daily blood hormone determinations and ultrasound scans of her ovaries, to see if follicles are visible. As the presumed date of ovulation approaches she may be given other hormones to boost follicular development. Clearly this is a fairly demanding programme of out-patient visits, which is likely to dominate one's life. Although the ultrasound examination requires a full bladder, which is stressful in itself, some women have said that seeing their follicles was a pleasure in itself, because they had never been sure before that they did actually ovulate. [9]

When the onset of ovulation is detected, or perhaps slightly before, the woman goes into hospital, for removal of eggs by minor surgery under general anaesthesia. There will be some residual uncertainty as to whether any eggs will be obtained. They are so small that they can get lost during the procedure. When the laparoscope is inserted ovulation may already have occurred, particularly if the medical team does not work on a 24-hour basis. Depending on the clinic and the man's own feelings the semen may have been produced and frozen earlier, or the man may be asked to produce a sample around the time of the laparoscopy. Some centres seem totally to ignore the possibility that he might feel embarrassed at having to masturbate to order in semi-privacy. It is not unknown for men to become temporarily impotent when faced with this psychological challenge. Since their wives or partners are likely to be recovering from the anaesthetic they may be reluctant or unable to help. One enlightened Australian programme gets men to practise beforehand, by imagining the stress they may feel, so that they learn to deal with it.

Some hours after ovulation has been detected chemically several eggs are removed using a laparoscope. A thin plastic tube is pushed into the Graafian follicles on the surface of the ovary, and the egg sucked out down the tube, using a small pump. The egg is located with a microscope and placed in a culture medium. Anywhere between 1 and 6 or even more eggs may be recovered in this way. These can then be fertilised outside the woman's body. If none are obtained, then another attempt must be made at the next ovulation. Usually the eggs are left for several hours in the medium to allow them to complete their maturity.

This form of fertilisation is usually referred to as in vitro fertilisation. In vitro means in glass, and biologists often use the term to refer to processes made to occur apart from an organism, most probably in laboratory glassware. The opposite is in vivo, which means as it occurs within a living organism. The replacement of a fertilised egg, or developing embryo, within the female body is referred to as embryo transfer. [10]

After the eggs and sperm have been brought together in the culture dish, there is a wait of a day or two for the embryos to develop. They are examined very gently under a microscope from time to time. Some may stop dividing, and some may grow too slowly, in which case they are discarded, but in general this fertilisation step is now successfully achieved.

Somewhere around 36 hours later, when the embryo has reached the 2- or 4-cell stage, it is transferred to the uterus. It is gently removed from its culture medium and drawn up into a thin plastic tube. This is passed through the cervix into the uterus and the embryo left there, having been gently expelled from the tube. The transfer procedure is very quick. Women are often given tranquillisers to help them relax and the husband may be present, since some see this, rather than the earlier procedure, as analogous to conception.

Then the wait begins to see if a pregnancy has been established. This involves repeated pregnancy tests, up to the date at which the next period would be expected. If it occurs then another attempt has failed and this latest loss must be mourned. If it does not there is still a considerable chance of miscarriage, particularly in the early months. Despite the stress, some women have said that they find the time in hospital memorable, because they may be treated as special patients and they are likely to meet other women going through the same ordeal.[11]

There are at least three more sources of stress with which couples will probably have to cope.[12] One relates to the variations on the basic procedure that I mentioned earlier. There may be decisions relating to the use of 'surplus' embryos, if the doctors regard the implantation of more than three as unacceptably risky. It may be necessary to use donor semen. Secondly, some people are anxious about what they will tell their children, even though as the procedure becomes more widely used, this should be less of a problem. If some kind of donation of gametes or embryos is involved, then this resembles the problem with more straightforward forms of AID.

Lastly, there is the question of research and clinical experiment-ation. All aspects of ivf are being researched at present, so that most people involved in these programmes will be asked to act as an experimental subject. Whilst consent may be willingly given, and some people may feel they are themselves making a contribution to research that will help others, nonetheless, it must complicate the experience. For example, it must be hard to refuse such a request from a doctor on whom such high hopes are pinned. This is precisely the kind of issue that needs to be explored retrospectively, by someone who is not seen as part of the clinical team, in order that people speak freely and frank comments be fed back to the programme. It is certainly interesting, and not unexpected, that lay people who have been involved with ivf report such positive attitudes to research, including research on human embryos.[13]

Defining the success rate of this procedure is not at all straight-forward. We come back to this question in Chapter 5. At the moment the chances of a baby being born from an attempt at embryo transfer are about 1 in 10, for the most experienced medical teams, if one embryo is transferred. If three are transferred, then the chance of a successful pregnancy, with the birth of at least one child, rises to about 30 per cent.[14] So many women and their partners will face the prospect of going through this procedure several times, if their desire for a child is to be realised, and many will be disappointed even then. Many, too, will learn of the procedure but not be accepted on to an ivf programme, because the problems are too severe or the waiting lists too long.

I have deliberately tried to link together some of the technicalities with the experiences of the two people principally involved. It is particularly important not to lose sight of this, whatever view one takes of ivf, given the considerable psychological strains created by involuntary infertility, the use of ivf as a 'last resort', its experimental status and the considerable media attention that certain issues have received. At the same time the procedure involves other people too, and the feelings and plans of the doctors and scientists involved with ivf are also significant. This is why I discuss the development of ivf in the next chapter.

Steps forward

It is also important not to think that ivf 'solves' the problem of infertility. There are many other related medical issues that deserve our attention. Not every infertile couple either needs or wants extensive medical investigation. Often what would be most helpful would be the support of friends and some professional counselling to help them come to terms with their sense of loss.

Nonetheless, it is a shock to find that infertility services in the UK are so unevenly organised and that so little data on their use is available. A recent national survey by the Family Planning Association showed that out of 192 District Health Authorities in Britain, 105 make no special provision for subfertility investigations within the family planning services and only 16 indicated that assistance was available in local hospitals.[15] In many parts of the country, then, people have to travel long distances to answer questions about how often and in what position they have sex. Infertility specialists regularly discover that people coming to them have been very poorly advised, often in private consultations. This is a scandal, and something that ivf itself leaves untouched.

50

More generally in society we fail the infertile by not noticing their medical and psychological needs and by frequent insensitivity, even when meaning to do good. An extreme example may illustrate the point. After her most recent miscarriage a woman found herself in a hospital bed next to a 19-year-old who had just had her eighth abortion. When she asked why she had been put in such a position, she was told that she was there to punish the younger woman, who had had the abortion. Not only did no one ask the infertile woman how she might feel about being used in this way, but none of the self-appointed guardians of the public morality here asked themselves whether such crude and obvious gestures would do anything other than deepen the alienation of the young woman who was regulating her fertility in such an unfortunate way. For both of them their real needs had been ignored. There are now self-help and pressure groups concerned with infertility, including the National Association for the Childless, Child, and the British Organisation of Non-Parents.[16] It is vital for people planning improved health care in this field to listen to what they have to say.

Greater effort to prevent infertility is also vital, even though the causes are not well understood. In women one source is infection. Pelvic inflammatory disease (PID) is the name given to infections of the upper reproductive tract, that usually have spread upwards from the cervix. They often cause considerable pain and inflammation, but they can go unnoticed. In the US around 850,000 cases of PID occurred between 1973 and 1976, a quarter of these involving admission to hospital.[17] The commonest cause of PID is infection by micro-organisms that are transmitted sexually. Perhaps 50 per cent of cases are caused by gonorrhoea and around 25 per cent by another micro-organism, *Clamydia trachomatis*, although these figures seem to vary from country to country. Infection is also possible from childbirth, from an abortion, from a routine gynaecological investigation or abdominal surgery. Around 15 to 20 per cent of women with PID become infertile, through damage to their Fallopian tubes.

For both men and women greater awareness that fertility is not inevitable and can be jeopardised is very important. Men particularly often believe that there are no great risks in ignoring the occasional medical complications of sexual activity. Yet their sense of self-worth is often profoundly altered when it turns out that they are infertile. Greater responsibility earlier in attending to minor symptoms could save a great deal of misery.

Ivf is valuable. It is also morally and politically problematic and will remain so for a long while. The fundamental problem with it is not the facilitation of complex forms of parenthood, or the dependence on embryo research, which we consider in Chapter 5, but its effects as medical drama on people's self-confidence, self-esteem and sense of responsibility for themselves. Ivf is quintessentially 'reproductive technology', in the critical sense that feminist writers intend. It requires women and men to qualify as potential beneficiaries and for women to submit themselves to the rigours of medical investigation and the unknown hazards of experimental medicine, in the hope of becoming pregnant. They enter into the drama of ivf, so that their endocrine system can be mastered, and so that oocyte removal, fertilisation and embryo transfer may succeed. We may think of taking home a baby simply as a technical success. That is the danger. For regardless of outcome it is always also a test of the emotional strength of the woman concerned and of those giving her support. It is not just technical virtuosity that makes it work, to the limited extent that it does but also people's ability to draw on psychological resources to sustain them in dealing with the experience of the technique. If they come to appreciate their own strength, then ivf is vindicated. But all too often hospital medicine denies people that self-knowledge, since all decision-making is taken from them.

Chapter 4

In vitro fertilisation:
from animal agriculture to medical venture capital

Historical background

Where has ivf come from? Its roots are older than one might expect, reaching back into the nineteenth century, with important work done in the 1930s and 1940s.[1] What stands out from the history is the importance of work on agricultural animals and the support from other lines of investigation, including that on contraception. Ivf was a research technique, as much as anything, something that researchers sought to achieve in order to understand better the phenomena of fertilisation, development and heredity. It only really came to have the purpose we now associate with it, the alleviation of infertility, in the 1960s.

It was a rather unfashionable topic, picked up and pursued by a few people, either in veterinary science or in medicine, where they took something of a professional risk with it. A long history of sparsely funded, spasmodically controversial research has given way to a period of very rapid expansion. In considering how other medical innovations may spread we should keep the form of this very rapid transition in mind. It is not implausible to suggest, for example, that something similar might happen with cloning in the 1990s.

One early pioneer was the Cambridge physiologist, Walter Heape. He wanted to bring scientific research to the animal breeding industry and worked for a while on the use of artificial insemination in dog and horse breeding.[2] But Heape is best known today as the first person to carry out the successful transfer of embryos, in rabbits.[3] His purpose was to investigate the doctrine of 'telegony', or the widespread belief that the effects on the uterus of bearing a particular set of offspring were transmissible to later progeny from the same animal. The phenomenon was of great interest to animal breeders, who believed that their prize females could be 'spoilt' by unintended coupling with an inferior male.

Heape removed two embryos from an Angora doe rabbit, which had been mated with an Angora buck, and placed them in the Fallopian tube of a Belgian Hare doe rabbit, which had been mated with a buck of its species. The Belgian Hare gave birth to six young, four of which resembled her and her mate, and two were undoubted Angoras. Heape felt that his experiment proved the feasibility of the technique of embryo transfer and showed no evidence that the uterus of the foster-mother had affected the Angora young. It is important to remember here that fertilisation occurred inside each animal, and the embryos from one were then removed on the tip of a needle, and placed in the other.

The turn of the century saw the appearance of the new concept of hormones, chemical substances released into the bloodstream by particular glands or tissues, so as to produce a very specific physiological effect. Appealing though this idea was to many physiologists, the problem was to find a reliable way of demonstrating what effects hormones had. They were particularly critical of the rather loose way that doctors approached this problem, with the hope that the extract of some organ, say the testis, would be useful in therapy.[4]

By the mid-1920s, however, various researchers had shown that the ovary not only produced eggs, but also released hormones during the menstrual cycle.[5] Gradually more and more sex hormones were isolated and their role in the cyclic production of eggs, in altering the condition of the uterus and initating pregnancy elucidated. Great hopes were built up that such findings could be applied to the regulation of fertility and the alleviation of what were seen as psychosexual disorders causing marital instability. The 1920s and 1930s were a period of great confidence in the possibility of engineering a society of stable, planned, nuclear families in this way. In 1928 a fertilised human ovum was identified in the Fallopian tube and a more definite estimate given of the moment of ovulation in the menstrual cycle.[6] Even so precise determination of that event has remained a problem for many years.

·

54

In the 1930s the American biologist, Gregory Pincus, discovered that oocytes removed from their follicles would continue their maturation. Released from whatever control had held them back in the ovary, outside the body they began to sort themselves out for fertilisation. One could thus obtain any number of ova for experimental purposes by surgery on the ovary. Maturation and fertilisation could then be studied more easily. By 1934 Pincus claimed to have fertilised a rabbit egg in vitro. Even though he later achieved international renown for his work on an oral contraceptive, some of Pincus's experiments from this period remain controversial. It is therefore questionable that he had actually achieved fertilisation here, since it is quite possible for eggs to begin to divide spontaneously, though not for long.[7].

Nonetheless, in the late 1930s the possibility of ivf being used to alleviate infertility was canvassed in a major medical journal.

Pincus and Enzmann have started one stage earlier with the rabbit, isolating an ovum, fertilizing it in a watch glass, and reimplanting it in a doe other than the one which furnished the egg, and thus have successfully inaugurated pregnancy in an unmated animal. If such an accomplishment with rabbits were to be duplicated in human beings, we should in the words of 'flaming youth', be 'going places'. The difficulty with human ova has been that those recovered from tubes have regressed beyond the possibility of fertilization in vitro. But by utilizing the electrical sign [of ovulation] we may be able to obtain them from the follicle at the peak of their maturity. If the new peritoneoscope can be developed along the lines of the operating cystoscope, laparotomy may even be dispensed with. What a boon for the barren women with closed tubes![8]

This, then, was forty years before the ideas mentioned here were realised in the birth of a baby. One of the scientists whose work provoked this editorial was the Harvard physician, John Rock, who ran an infertility clinic in a Boston hospital. From 1938 to 1952 he and his collaborator, Arthur Hertig, examined over a thousand Fallopian tubes and wombs from women having hysterectomies.[9] Their surgery was timed to occur just after ovulation, and the women were asked to record, whether they had had intercourse in the days before the operation. The hope was that in some cases fertilisation would have occurred and that the ovum could be discovered.

The first human embryo was found in 1938. Hertig carried it by hand on an airplane from Boston to the Carnegie Institution in Baltimore, which maintains a large collection of embryos of many species. In 14 years Rock and Hertig collected 34 fertilised eggs from 210 unknowingly pregnant women.[10]

At the same time Rock was also interested in attempting human in vitro fertilisation. Again women waiting for a hysterectomy were asked to allow their ovaries and tubes to be searched for unfertilised eggs. These were then examined and exposed to the sperm of interns paid by Rock. Much of the practical work was done by Miriam Menkin, Rock's assistant, who often had to sit for an entire day outside Rock's operating theatre waiting for the tissue to be sent out. After many unsuccessful attempts an egg finally divided into two after fertilisation in February 1944. In the excitement of the next few days the egg was lost. In the next few months she succeeded again three times and a paper describing the work was published in the prestigious journal *Science*.[11] The claim that this was authentic in vitro fertilisation is not now generally accepted. Nonetheless they attracted a great deal of public attention, some of it very critical. Neither in their 1944 paper nor in a later one in 1948 was any mention made of the possibility of re-implanting this dividing cell. But they must have contemplated the prospect. The idea was apparently suggested by women writing to Rock after the research had been made public.

Whatever they had in mind, the idea of transplanting embryos was being tried with agricultural and laboratory animals in the 1940s and early 1950s. But these were embryos created in vivo, by normal mating, as Heape's had been, or artificial insemination, and then flushed out of, or surgically removed from, the body of one animal to be replaced in another. By the 1950s this could be done with cattle, with difficulty. In 1959 Chang published a paper on his work with rabbit eggs which is generally taken as the first authentic achievement of in vitro fertilisation.[12]

To someone trained in animal genetics and reproductive biology in the early 1950s these results would have been interesting, indicating that the phenomena of fertilisation could be explored in the laboratory. But it was the development of cytogenetics – the study of the elements in the cell like chromosomes involved in the process of inheritance – that especially stimulated a young British biologist, Robert Edwards, to study the maturation of oocytes. In the mid-1950s techniques were created that allowed chromosomes to be made visible and studied more easily.[13] Chromosome re-arrangements in maturing eggs can go wrong. The prospect of being able to understand genetic abnormalities arising early in development appealed to him. It was 1965, however, by which time he had moved to Cambridge, before he published details of the maturation of human oocytes, observed outside the body. The possible value of this research for the alleviation of infertility was discussed in his article in *The Lancet* in November 1965.[14] But he could only get small numbers of eggs from co-operative surgeons.

The decisive step forward came with his reading of the technique of laparoscopy and contacting its pioneer in Britain, the Oldham gynaecologist, Patrick Steptoe. He had pioneered the uses of the laparoscope, a flexible viewing instrument, much improved in France and Germany in the 1950s, and could claim to be one of the very few people in the world with the dexterity and clinical skills needed to use the instrument. Such was his commitment to developing his expertise with it that, with a heavy case-load at Oldham General Hospital, he practised on cadavers in the hospital mortuary at lunch-time.

There was some scepticism amongst doctors about the value of laparoscopy in the early 1960s. The abdomen had to be distended with gas to separate the internal organs and some people thought that risky. Then there was the chance that you might puncture something as the laparoscope was inserted. Moreover, the early instruments used a light source that gave off heat, which meant they had to be used very quickly. But fibre optics, which allowed light to be transmitted into the body from an external source, removed that problem. Finally, there was the novelty and difficulty of recognising things through what was in effect a small, bendy telescope under new conditions of illumination. Sizes, textures and colours all had to be judged anew and that took time to learn. So not everyone thought it useful. I am reminded of the struggles of the seventeeth-century astronomers to persuade sceptics that the telescope really did improve one's perception of celestial bodies. The novelty of the technique of laparoscopy, the slowness of its uptake by other doctors and Edwards's desire to press ahead was such that he was prepared to contemplate lengthy trips to Oldham where the laparoscopy was to be performed, to remove oocytes from some of Steptoe's infertility patients.

By February 1969, when they published a paper in the leading scientific journal *Nature*, they had worked with 73 eggs.[15] Of these, 18 could be seen to have been penetrated by a sperm, and some had also gone through the preliminary stages of post-fertilisation cell division. Accordingly they felt able to claim that most of these 18 eggs had probably been fertilised. They called a press conference to discuss these results. By the following year they were able to get eggs to divide to the blastocyst stage.[16] This kind of result was much harder to doubt and their work became the subject of considerable public concern.

These early papers were published in widely read scientific and medical journals. They were well publicised in the mass media and some of their possible implications considered there. Edwards and Steptoe both began to publish their thoughts on a few of the moral issues their work could be said to raise. They pressed ahead with their clinical experiments, using volunteer patients in Oldham. They also approached the Medical Reseach Council for funds to start a research institute but were rebuffed. Elsewhere various physicians began themselves to think of trying in vitro fertilisation, amongst them Pierre Soupart in the United States and Carl Wood in Australia, to whom the Ford Foundation began to give substantial backing during the 1970s.

In the spring of 1972, Dr William Sweeney, an obstetrician at the New York Hospital Cornell Medical Center, made contact with Dr Landrum Shettles, who was then at the Columbia-Presbyterian Hospital Medical Center, about a patient, Mrs Doris DelZio, on whom Dr Sweeney had already operated unsuccessfully on two occasions to repair her Fallopian tubes.[17] Dr Shettles was known to have a long-standing interest in ivf and he agreed to attempt in vitro fertilisation of ova from Mrs DelZio with her husband's sperm. This was done, on a trial basis, without subsequent re-implantation, on June 6 1972.

A second attempt was made in September 1973. The following afternoon Dr Shettles was called to the office of his superior, Dr Raymond Vande Wiele. There he discovered that the ovum in its culture medium had been deliberately exposed to the air so that it had ceased to develop. Dr Vande Wiele said this had been done because it was felt that Dr Shettles was insufficiently skilled to attempt this work. Eventually Mrs DelZio and her husband brought an action for damages against Dr Vande Wiele, the Presbyterian Hospital and Columbia University, and were awarded $50,003.

Edwards and Steptoe were making attempts to re-implant embryos at around this time. In Australia an attempt at implantation was made in 1973. On September 29, two weeks after the events in New York, *The Lancet* carried a report of an unsuccessful implantation of an 8-cell embryo by Carl Wood's group.[18] That putative pregnancy ended after 9 days when the abdominal wound ruptured. These events were not public knowledge, but they suggest a context for the actions of the physicians in New York. The DelZio case finally reached the courts in 1978, just as Edwards' and Steptoe's first success became known.

In July 1974 another remarkable event occurred. Dr Douglas Bevis, professor of obstetrics and gynaecology at the University of Leeds, announced at the annual meeting of the British Medical Association that to his knowledge 3 children born after ivf were alive and well in Britain and Western Europe[19] No details of this have appeared subsequently, although a colleague in Leeds confirmed that these births had occurred, and the whole episode remains a puzzle.

The attempt at ivf which led to the first successful birth was carried out on November 10 1977.[20] On 18 April Steptoe revealed in a telephone conversation with the lawyer handling the DelZio's suit that they had reason to believe that the first woman in their programme was about to pass the sixth month barrier, at or before which all their previous pregnancies had failed. This set in train a massive media exercise to try to get details of this special pregnancy. By July it became clear, that after something like an auction, an agreement had been reached with the Associated Newspaper Group, which publishes the *Daily Mail*, guaranteeing exclusive rights to the story of the couple involved for a sum said to have been £315,000. At that time this was one of the largest cash payments ever made by Fleet Street for an exclusive story. It only intensified the frenzy of largely useless journalistic activity in Oldham.

On July 27 every national newspaper in Britain carried some report of the birth of a baby girl late on July 25. Reports dealt with the technical details of the procedure, offered profiles of Edwards and Steptoe, but made no reference to any of their colleagues or assistants. In the *Daily Mail*, which devoted nine pages to the story, pride of place was given to photos of the baby and her delighted, grateful parents. Though individual countries and cities have from time to time celebrated 'their' first baby, nothing like the excitement over the world's first has been seen again. When the second baby in Britain was born in January 1979 it was scarcely news at all.

In all three babies were born from the Oldham-based programme. With Steptoe's retirement from the National Health Service in the summer of 1978, other possibilities had to be explored if their work was to continue. By 1980 a clinic had been established at Bourn Hall, just outside Cambridge. This minor stately home was converted into a small hospital. Initially the Associated Newspapers Group was said to have had some financial involvement, but in 1980 the *Guardian* newspaper reported that Edwards and Steptoe had bought the house themselves, along with Alan Dexter, an accountant and former hospital administrator. They now operate a training scheme for doctors, as well as assisting patients from all over the world. Two of their junior colleagues have just left to join another private ivf clinic in the UK.

Five years after the Oldham success Clifford Grobstein wrote an article evaluating progress up to that point. His literature review covered the work of 11 groups, including 4 in Australia, one of which was led by Carl Wood, 1 in West Germany, 2 in France, 2 in the United States, one of which involved Edwards' former collaborator, Howard Jones, and 1 in Britain, Edwards and Steptoe.[21] This total was an underestimate, caused understandably by listing only those groups which had actually published results. Together these groups had reported in the literature 1,110 embryo transfers, leading to 184 pregnancies and a smaller, unstated number of live births, probably around 100.

In the summer of 1984 the Third World Congress of In Vitro Fertilisation and Embryo Transfer was held in Helsinki, with contributions from physicians from 25 countries. A questionnaire discussed at the conference had been sent to 200 separate groups around the world, all offering ivf. Between 1978 and January 31 1984 590 births had resulted from 517 well-established pregnancies and a further 570 pregnancies were still continuing.[22] By now over 1,000 children will have been born after ivf around the world. The real expansion of the technique occurred in 1983 and 1984, as the doubts about the risks and the likelihood of success have subsided in doctors' minds.

The rate of progress in various countries has been strikingly different. In Britain and Australia research has gone ahead without major opposition, if one discounts the lack of support from the MRC in the UK and the occasional harassment of ivf centres by anti-abortion groups, trying to establish grounds for legal action against research using embryos. As one might expect in the predominantly Catholic countries of Western Europe, things have developed rather more slowly. In Japan, where attitudes to abortion are very tolerant, ivf has been developed by several groups.

But in the United States, where there was an active group of researchers in the early 1970s, and, as we have seen, one can trace that interest back to the 1930s, work has been surprisingly slow, until very recently. In part this was due to a hostility to the idea of foetal research, which came to include any work with embryos, as a reaction to the shift in the law relating to abortion, after its legalisation in the United States by the Supreme Court in 1973. Because of the highly charged political environment, following that decision, the National Institute of Health, the agency that provides the bulk of the funds for biomedical research, announced that all research proposals involving ivf would have to be assessed by a specially created National Ethics Advisory Board. Several years went by before that body even met, which effectively imposed a moratorium on American researchers. The Board held public hearings and commissioned a series of detailed studies in 1978, and then issued a set of recommendations that would have allowed ivf to be investigated, but these suggestions were ignored by successive Secretaries of Health, Education and Welfare – the government minister with responsibility for the NIH.[23]

In January 1984 the American Fertility Society issued an 'Ethical Statement on In Vitro Fertilisation' in its journal *Fertility and Sterility*, which was an important signal that doctors in the US felt ivf to be acceptable. Later that year, stimulated by the example of the British government inquiry, Congressional hearings were convened to review the situation. But as in Britain now there must be a real prospect of a reaction against ivf, led by anti-abortion lobbies and those prepared to ally themselves with them, even if not so powerfully exercised by abortion itself. If the legal, moral and social issues had been considered formally at an earlier stage than was the case in Britain at least, and there was no practical reason why this should not have happened, then the prospects of the continuation of ivf might not be so vulnerable to the upsurge of reactionary sentiment, as they are today.

In Britain the political nettle was finally grasped by various august bodies, the Medical Research Council, the British Medical Association, the Royal College of Obstetricians and Gynaecologists and the Council for Science and Society around 1981. The government itself solicited advice from a committee of inquiry, peopled by the Great and the Good. It was led by the Oxford philosopher, Mary (now Lady) Warnock, who had previously chaired a committee on provision for educationally subnormal children. She was also a bridesmaid at the present Prime Minister's wedding.

Amongst many other recommendations, her committee endorsed the continuation of ivf in the UK and recommended that a limit of 14 days be set beyond which experimentation with human embryos, created by ivf, should not be permitted. That limit has been applied by a Voluntary Licensing Authority, sponsored by the MRC and chaired by Lady Frances Donaldson, which oversees work in this area for the time being.

Since the appearance of the Warnock report, the volume of public discussion about ivf in the media has increased markedly, as has parliamentary interest. Enoch Powell, MP, presented a private member's bill to the House of Commons in 1985. This sought to ban all research on human embryos, and would have required notification to the Minister of Health of every attempt to implant a human embryo. Furthermore all embryos would have had to be implanted. This draft legislation failed to reach the statute book in the early summer of 1985. Similar proposals were brought forward in the next session of Parliament, but also failed.[24]

Elsewhere in Europe expert committees of various kinds have been set up to consider ivf and related issues.[25] In general they have endorsed ivf, and in so doing have been somewhat in advance of public opinion. In the Netherlands limited provisions were announced in 1985 to allow the reimbursement of the costs of ivf through the state medical insurance scheme.[26]

However, the most thorough discussions of ivf and related procedures have taken place in Australia. In the state of Victoria a committee of inquiry, chaired by Professor Louis Waller, met from 1982 to 1984. The recommendations of his committee were very quickly adopted as the basis for legislation as the Infertility (Medical Procedures) Act.[27] This instituted a system of control over who is permitted to perform ivf, where and within what limits. It also restricts ivf to married couples. The following year a committee established by the Family Law Council, which advises the Federal Attorney-General, produced its own report, and listed some thirty discussion papers and reports from various state bodies throughout Australia.[28] This represents the most conscientious formal attempt to think about ivf anywhere in the world.

However, Australia has also seen an important controversy over the privatisation and commercialisation of ivf. In the autumn of 1984 academics at Monash University discovered and publicised plans to create a company, backed by venture capital, that would sell the expertise of Professor Carl Wood and his team. This was to be called IVF Australia and was to operate in the US as well as in Australia. The financial backing was to come from a syndicate of Australian investors, organised by Mr Robert Moses, an American citizen who was to be managing director. He and his wife had in fact been involved with Professor Wood's programme.[29]

Quite the most remarkable aspect of this story was the secrecy surrounding the financial basis of the venture, and the returns to Carl Wood and his colleagues. Critics of the project were particularly concerned about the clear indication that the investors wished to control disclosure of scientific and technical information originating from the ivf programme for commercial reasons. Not only did this suggest that novel techniques and results might well be patented, but it would also make the assessment of all the morally problematic aspects of the work even harder. As Reverend Father Francis Harman, a member of the Waller Committee, put it:

> [M]y understanding of the representations made by all the protagonists of the technique was that the IVF quest sprang from concern, the IVF research was motivated by idealism, and the IVF practice was altruistic. Perhaps the more mundane reality now betrayed in the association with 'unidentified business interests' should alert the community to the complications . . .[30]

Chapter 5

Evaluating the criticisms of ivf

Ivf and the sanctity of life

Despite the momentum behind it, after an explosive period of growth in the last three years, we must ask if ivf should continue, and if so, on what basis. In essence this is what the intense recent debate concerning research with human embryos has been about.

Some people object to the very idea of enacting fertilisation outside the body, without intercourse. For them ivf rends asunder processes and feelings that should form one divinely ordained whole. Ivf is not only unnatural but also contrary to God's purpose. I cannot accept either proposition. To say that something is 'contrary to nature' is virtually incomprehensible in the medical domain. If one obeyed this rule generally, we would not interfere with the 'natural' course of bacterial infection or the 'natural' loss of blood after an accident or the 'natural' growth of a tumour. The desire to regard procreation as special, even something sacred is more easily understood, but not why this should stop us from altering the form or the context of the process of human fertilisation, particularly if our purpose is to enable someone to have a child.[1]

A much more subtle and more widely shared objection to ivf concerns the sanctity of life and the supposed rights of human embryos to continued existence from the first moment of life. You can believe this and not be totally opposed to ivf. But several aspects of the basic procedure and associated research either require embryos to be destroyed or place their continued development in jeopardy. On this view ivf, then, should only be practised in ways that allow any embryo to take its chances in the survival course of implantation and pregnancy. The underlying motivation is to treat early embryos as if they were human beings with the same rights as fully developed members of the human species. Hence the recent attempts in the UK, as in the Unborn Child (Protection) Bill of 1985, not to outlaw ivf, but to prohibit research with human embryos and to require the implantation of all fertilised eggs.

In Chapter 3 I detailed the stages of fertilisation and the contingency of early embryonic development. Three points stand out clearly from that picture. Firstly, there is no one moment of fertilisation. There is no point that we can label unambiguously as the 'beginning of life', although there are any number of events that could be identified as essential to the formation of an embryo with the capacity for continued development. Secondly, the very idea of 'life' beginning within the process of fertilisation is a profound distortion, if not simply wrong. For both spermatozoa and ova are clearly alive. They combine at fertilisation to form another living system with a different kind of organisation. Fertilisation is a transformation, and truly remarkable one, but, if we care about how we use language, we cannot call it a beginning. Thirdly, what is created at fertilisation is an organised biological system containing a unique set of genetic instructions. It is thus individual in the loose sense of being distinct from any other cell-system existing just at that moment. It also belongs to the human species, because it is a combination of human germ cells. But it would be wrong to regard it as an individual human being because it has their potential to divide into twins or to become a tumour of the placental tissues, called a hydaditiform mole. We cannot say that the existence of a specific human being has begun, only that that is the most probable outcome if the embryo survives.

Early embryos are thus a very remarkable kind of entity. They are in a state of very rapid internal re-organisation. Their cells are organising themselves into layers. Some cells will give rise to the developing organism and some to the membranes that surround it in pregnancy. The question is what moral status does organised matter in this form have? In particular do early human embryos have a right to continued existence? Is it their potential to develop into an independent human being that is important or the fact that they acquire new characteristics and functions as they develop? Embryos have intrinsic qualities that should command our respect – complexity, capacity for self-organisation, robustness and so on. We diminish ourselves if we treat or use them thoughtlessly. But I cannot see that the argument based on the potential has any force, imposing upon us the obligation to allow them to develop, if at all possible. It is what they have become at any given stage, not what they have the possibility of becoming, that is morally significant.

Our approach to rights in other areas is also developmental. The right to own property, the right to hold a passport, the right to receive education and the right to vote are all recognised or conferred at specific stages. It is quite legitimate to take the same approach during embryological development, and the laws in many countries regulating abortion are framed on this basis, the right to protection from medical procedures which would terminate the existence of the foetus being recognised with the acquisition of the capacity to live outside the womb. Research with human embryos is a more complicated issue which I take up later. My point here is we cannot say that ivf is invalidated because early human embryos have a right to life. They have no such rights. The curious thing is why some people should have wanted misguidedly to be such impassioned champions of this category of living beings in the mid-1980s.[2] It is much more important to consider the rather less familiar biological and medical difficulties with ivf.

The creation of abnormalities

Human embryos are remarkably small, given that they contain within themselves such a richness of developmental possibility. In the seventeenth and eighteenth centuries the idea that they could be so minute was almost unthinkable. It was not until this century that one was actually found and observed. But we now have the confidence to fertilise ova away from the warm, sticky interior of the human body and then transfer them back as they begin to develop. In the early 1970s it was commonly asked how anyone could be sure that abnormalities would not be produced by handling early embryos or by some subtle environmental nudge to the fertilisation process. Since no one could do more than make informed guesses, some scientists argued that it would be better to wait until more evidence was available, particularly from more experimental studies with animals.[3] Not only was there the prospect of gross abnormalities, which would probably be detectable in mid-pregnancy, but it was also not impossible that in vitro fertilisation could be the cause of some subtle medical problem, that would only appear after birth.

That, for example, was the position of the Medical Research Council in the UK. Edwards and Steptoe did not share it, nor did the Australian researchers, because of doubts of the value of extrapolating from mice or rabbits to the human case.[4] Although I cannot document this, I suspect the majority view amongst obstetricians around the world in the late 1970s was that the risks were non-trivial. They held back from getting involved themselves. As we know, some people went ahead. By 1982–3 sufficient children had been born without any of the postulated major defects to vindicate their confidence and dozens of newcomers then entered the field.

An important argument for pressing ahead was that studies have shown that a very high proportion of human embryos conceived within the body actually have chromosome abnormalities. Of these a very high proportion are eliminated precisely for that reason. Either they fail to implant or the pregnancy does not continue, being compromised in some way by the chromosome defect. Evolution has created, it was argued, a very efficient mechanism for rejecting defective human embryos, upon which one could rely even if fertilisation had taken place outside the body.[5] It was also argued that pre-implantation embryos are fairly tough and they can tolerate a degree of environmental assault. They become more sensitive a little later in development.

It is one thing to state this argument, another to test it. For example, you have to construct a control group of 'normal' abnormalities to compare with those arising from ivf. To do this you need data from several thousand pregnancies.[6] In other words, on the assumption that developmental abnormalities will be fairly rare, one begins to do ivf and to accumulate comparative data.

The consensus of ivf workers in mid-1984 was that even though a very few children have been born with congenital and chromosome defects, the risks of abnormality are no greater than with normal fertilisation.[7] The debate about risk is not settled. For example it will soon be possible to use for ivf sperm from men who produce very few spermatozoa, of which a high proportion may be 'abnormal'. By micro-injection into the egg the chances of fertilisation can be considerably increased. But is this safe?

There may possibly be less obvious, less immediate medical problems in ivf children, of whom the oldest are only 5 or 6 now, so it is clearly important to gather some data on such children. Since one wants to compare their health with that of their parents, one needs some medical data about them as well. In Britain, the United States and Australia registers have been established to collect such medical information.[8] This in itself may not be easy. Psychologically it is important that such investigations keep a low profile, yet if they are to be useful the information needs to be complete. So little of modern medicine is evaluated that it is vital to see this is done for ivf.

Risks to women

Whilst some attention has been paid to the risks to children from intervention early in development, very few people have spent much time on the possible risks to women. Ivf commonly involves large doses of hormones and the repeated rupture of ovarian follicles. Some of these drugs are known to interfere with the endocrine function of the ovary, and their long-term effects are largely unknown. Gena Corea has argued strongly that these risks have been given insufficient attention.[9] Multiple implantation of embryos also increases the risk of multiple pregnancies, which are more likely to involve complications. Certainly it would not be the first time that a misplaced sense of obligation allowed men to take risks with women's bodies. At the very least doctors have a duty to organise the most careful follow-up studies of women in ivf programmes.

There is at least one case of a fatality involving ivf, although a special case, but a revealing one nevertheless. A group of Brazilian doctors invited a specialist from one of the well-known ivf groups to perform a demonstration of the technique, which was to be televised. During the oocyte recovery the woman died under the anaesthetic, and the exercise was abandoned. Anaesthesia always involves a slight risk, but possibly in the excitement of the event a mistake was made, which cost the woman her life. The episode exemplifies the possible problems of the very rapid diffusion of a new procedure, driven by medical and commercial enthusiasm.

Whilst there is no strikingly obvious evidence that ivf is particularly hazardous, and not enough to require a moratorium, much more information on long-term effects and unexpected complications needs to be compiled and its significance considered publicly. Since the 1960s women's organisations have begun to push much harder for an open assessment of technologies and procedures that impinge directly upon their health. This is particularly true in obstetrics and gynaecology where women have argued successfully for detailed studies of the safety of oral contraceptive pills and injectable contraceptives, the induction of labour, and foetal monitoring, in the face of complacency and hostility from the medical profession. At the very least there is a case for trying to evolve procedures which make less use of large doses of hormones and less use of anaesthesia, and for involving women in the assessment of these risks. It is no argument to say that infertile women are often prepared to accept significant risks and discomfort in order to conceive. Even if that is true, they still have a right to the best-informed advice as to the known problems and the many uncertainties that remain.

Reasonable failure rates and natural comparisons

What is an acceptable risk is linked to the question of what is an acceptable rate of failure. Conventionally the statistics are described as success rates. They are often stated incompletely and sometimes misleadingly. A recent editorial in *Fertility and Sterility*, where many ivf groups publish their results, sharply reprimanded those who have been a little vague with their data, to put it most charitably, in order to bid for custom in an increasingly competitive market.[10]

Overall several basic points stand out. Firstly, success rates have risen from the late 1970s, although they remain vulnerable to environmental factors like a change in the water supply or new laboratory supplies.[11] In their Oldham work Edwards and Steptoe achieved 3 live births amongst a group of 77 women. One *could* say that their success rate was 4 per cent. They and many other centres now do much better, although each new group has to go through a learning period. Secondly, there are some marked variations between the groups, but a world average can be computed, which is a kind of benchmark. However, quoting this figure out of context can be very misleading. It is vital, for example, that couples are told what success rate a particular clinic is achieving. Thirdly, and most importantly, there is a difference between the probability of eventual success and that of a live birth after ivf in a given menstrual cycle. Both these figures depend on how many embryos are transferred at each attempt. In any case what is a reasonable best figure to expect. Is 100 per cent a plausible goal? Or is something rather less than this the likely and acceptable limit?

Data from all over the world was compiled for the meeting in Helsinki in 1984, from 58 ivf teams. Together they had actively intervened in 9,641 ovulatory cycles and achieved 1,209 viable pregnancies, which is an overall success rate of 13 per cent per cycle.[12] Breaking these figures down further, we find that the current pregnancy rate is 10 per cent per cycle if one embryo is transferred, 15 per cent if two are transferred and 19 per cent with three. Some experienced groups have published data indicating that they do slightly better than this. It is advantageous to try to arrange the recovery of as many oocytes as possible, so as to have three embryos to implant. This requires treatment to induce superovulation, increases the chances of multiple pregnancy, and makes it more likely that 'surplus' embryos will be created.

If we take the average figure of 13 per cent per cycle, there is only a 40 per cent chance of success with 3 cycles of investigation. Some groups limit the number of attempts to 3. It is true that at some centres if 3 embryos are transferred then the chances of success are rather better. But for the foreseeable future more couples will be disappointed having gone through the procedure than will be rewarded with success. These figures underline the need for people to be prepared for failure. There is evidence that they often have unrealistic expectations of their chances of having a baby.[13]

But what is a 'reasonable' success rate? What is it for pregnancy based on intercourse? The French statistician Henri Leridon describes the statistics of conception and embryo survival thus.[14] Of 100 human ova exposed to spermatozoa about 16 will not be fertilised, and 15 more will not begin to divide. A further 27 will not implant successfully. By now we are down to 42 per cent of the original group. The surviving embryos are about two weeks old. The number of live births to be expected 36 weeks later is 31, indicating a success rate of 31 per cent. Some estimates put the loss in pregnancy a little higher, and the overall success rate at 25 per cent.[15]

Making a comparison between ivf and 'nature' is complicated by the fact that in the former only those embryos which have been fertilised, have begun to divide and are growing in an acceptable manner will be re-implanted. They are already a select group. The 'natural' control group is the 69 remaining ova in Leridon's sample which may or may not implant. Of these 31 survive (44 per cent). So one might say that the 'natural' success rate from the beginning of implantation onwards was 44 per cent and that this was a reasonable standard of comparison for ivf. This figure is open to question, but at least it indicates strongly that recurrent failure is intrinsic to the procedure. In a word, it is natural. The problem is that with ivf, as with strongly desired conception being sought in other ways, one cannot but help notice failure, and the routines of hospital visits, surgery and medical consultation create a dramatic context that increases the tension.

Deciding on an acceptable level of failure is difficult. Quite reasonably different standards can be applied to work that is clearly seen as very exploratory, and work that forms the basis of a routine programme. Whereas a few per cent would probably be acceptable in the former case, it would not in the latter. The primary issue is that people considering ivf are given accurate data, in a form that is easily comprehensible to them. The dangers are that the statistics can be compiled in ways that exaggerate actual performance or that some global average is substituted for up-to-date figures on the particular hospital team.

Infertility and the demand for ivf

We move now from matters that could be decisive objections to ones that are less critical but are still very awkward. Of these the most salient is the question of demand, and the impossibility of meeting it at present. Population statistics indicate the scale of the problem. I have used American data because it is more accessible and have omitted all methodological comment and qualification.

The years from 1940 to 1965 saw a 'baby boom' in a number of developed countries. Some couples were 'catching up' and some were not waiting as they would have done had they been of reproductive age in the Depression years. Population surveys from the 1950s show that very few couples sought to remain voluntarily childless. It is primarily the children of this baby boom generation, who began from the mid 1960s onwards to make widespread use of the new forms of contraception, whose infertility problems doctors are now seeking to treat. Exactly why involuntary infertility should have become more common is not known, although the increased incidence of sexually transmitted diseases, the complications of oral contraceptives and toxic chemicals in the environment have been discussed as possible causes.[16]

National fertility surveys from the mid 1960s onwards show increasing childlessness again, some of which is long-term and voluntary. Some writers claim that a life-style without children is now being freely chosen by more people, given the greater availability of reliable forms of contraception. Some people see themselves as 'child-free' rather than childless.[17] But it is more obvious that the period from the mid-1960s to the present has seen the deferment, not the abandonment, of childbearing.

One consequence is that more people will only discover later in life that they have difficulty in conceiving. They will then feel they have less time in which to do something about it. That in itself is one factor behind the increased demand for medical assistance with infertility problems. Also people from the baby-boom generation have generally tended to ask more from medicine than their parents did and to know that more is technically possible.

Data from a very recent US fertility survey of women, married and unmarried, aged 15 to 44, shows that 2.7 million women or their partners had or seemed to have some physical reason that was preventing them from conceiving.[18] Of these 1.6 million were childless. Another 2.4 million (or their partners) were surgically sterile for non-contraceptive reasons. Of these 0.2 million were childless. Altogether, then, that is 1.8 million childless women unable in their present medical and marital state to have the child they would have liked. If we count those with children already who would have liked more, the total is around 5 million. This is around 1 in 10 of all US women aged between 15 and 44. Despite these large numbers it is quite likely most fertile couples will have no friends or relatives who are having difficulties and will assume that fertility is inevitable. For their part people who are infertile may well feel that their problems are simply not appreciated by the fertile majority and thus never mention them, thus compounding their sense of isolation.

Clifford Grobstein and his colleagues have tried to estimate the likely demand for ivf using this same census data from the late 1970s in the United States.[19] They suggest that 850,000 women have damage to their Fallopian tubes, which makes pregnancy unlikely or impossible. Of these perhaps 40 per cent would like to have children, and of this group perhaps 15 per cent will conceive without ivf, leaving a total of 290,000 women. They also suggest that perhaps 150,000 women without Fallopian tubes would like to become pregnant and might be considered as 'candidates' for ivf. Together this comes to 440,000 women. This is about 2 per cent of couples in the US today in which the woman is of reproductive age. They assume that a 12-year period of investigation and waiting for ivf has to be anticipated. On that basis one-twelfth of this group could be dealt with every year – or 36,000 women. In fact the existence of a 'backlog' of women as things stand means that the situation is worse. Also, if one adds people seeking assistance because of other forms of infertility, then the demand is immediately increased again. Grobstein reckons it could be doubled to around 70,000 a year in the United States.

If we divide these figures by 4 to allow for the differences in the sizes of the national populations, we reach a figure of about 100,000 women with tubal infertility that could be alleviated by ivf in Britain, or 200,000, taking all causes into account. With a 12-year total waiting period that is 18,000 a year. Very recently Michael Hull has published data on the extent of infertility in the UK, based on a study of people attending his clinic in Bristol.[20] His conclusion is that about 1 in 6 couples will require specialist help with infertility at some time in their lives. Of those who seek help ivf could be of assistance for around 18 per cent. This represents a possible demand of 216 new couples per year per million of the general population. In the UK that would then be around 10,000 per year or roughly half the number arrived at with Grobstein's assumptions. In the distant future one centre per health region could possibly cope with this demand, although that would imply that about 12 women a week were being dealt with in each clinic, or 600 a year. At present the NHS-funded clinic at St Mary's Hospital in Manchester is dealing with 6 women a week and has a 4-year waiting list. But the problem at the moment is the size of the pool of people who have become infertile. Even if we make very conservative assumptions about who will want ivf and which couples will actually get referred to the clinics, demand clearly exceeds the supply of services. By October 1985 ivf was being done in 10 NHS hospitals in England, 5 of them being in London.[21]

Under these conditions some doctors have decided to exploit the commercial opportunities afforded by private practice. Steptoe's and Edwards' clinic in Cambridgeshire provided ivf for something like 1,200 women in its first two and a half years of operation. That is about 10 a week, although now they probably deal with more. By late 1985 there were 6 private hospitals or clinics offering ivf, for those with the money to pay. The cost of ivf in the UK is around £2,000 per attempt. Grobstein's estimate of the average total cost in the United States with several attempts to have a reasonable expectation of success is $35,000, a sum unlikely to be met by insurance schemes, because of the novelty of the procedure. At these prices only the well-to-do are going to have any hope of access to the private sector. This might just be acceptable as a kind of tax on the upper classes, but only if the technical and financial benefits are fed back into the State system, and if doctors can be made to concern themselves with more mundane and less profitable medical matters.

There is a case for developing ivf as a specialist service within the National Health Service, but only if the overall provision of resources for infertility is carefully analysed and publicly debated. Inevitably within publicly funded programmes there will be selection; both self-selection by couples of themselves and selection by those controlling access to this form of medical assistance. Doctors will pick out from those referred to them those who in their view are best able to be helped in this way and most deserving of the chance. Selection must be based on medical criteria and not on covert judgments about supposed fitness for parenthood made by doctors alone.

Variations

There is a basic technical form to ivf, which I described in Chapter 3. There is now a growing number of variations in procedure, which will both boost demand, by extending the range of applications, and add to the list of refractory moral and legal problems which have to be considered.[22] The practical and cultural significance of being able to externalise fertilisation in this way is hard to overstate.

Ovum donation

One obvious variation is the use of donor sperm to fertilise the ovum, most obviously because the husband or partner of the woman seeking ivf is infertile. If AID is acceptable, then using it with ivf creates no new problems. It is just possible that people involved with ivf might consider AID, when otherwise they would not do so. Consider a couple where the woman is infertile, whilst both are the carriers of the gene for a recessive genetic disease. Having wanted children so badly and endured so much to get ivf, one could imagine that they might be reluctant to terminate the pregnancy if the disease were diagnosed in the foetus. Using donor sperm from a man known not to be carrier would obviate that possibility.

More radically, ivf makes egg and embryo donation possible. Donated ova would most probably be fertilised with sperm from the husband or partner of the recipient, into whom they would be transferred. This has been done already. Implantation can occur only at a specific moment in the menstrual cycle, so that the cycles of recipient and donor have to be brought into synchrony, by the administration of hormones, or the embryo frozen and thawed out again at the right moment for the recipient.

In a sense this is the analogue or mirror-image of AID, with the same asymmetry of genetic connection with the child – the man being a genetic parent, and the woman not, even though she would bear and give birth to the child. Ovum donation must also be acceptable morally, although it is clearly more complicated than AID. It involves giving more. If the donor is involved herself with ivf, then she is donating something of far greater value to her, and that is much harder to obtain and to handle, than is 5 cc of sperm to a sperm donor. The egg she is donating could be fertilised with her own husband's sperm and re-implanted in her. It will only be given away if several of her own eggs have already been implanted in her. There are several potential problems here to consider.

Firstly, there is the question of anonymity. This may not be that easy to achieve, if both women are in the same hospital for several days. In one case reported so far the women were in fact in adjacent hospitals, neither was told the identity of the other, nor whether implantation was successful. I can see that keeping things secret could make some sense, but is it right? If women feel that they would like to know what is happening in the other pregnancy, surely they have a right to be told? Secondly, there is the question of legitimacy. As with AID the genetic parents of the child are not married to one another, even though the future social mother bears and gives birth to the child. If the notion of illegitimacy were abolished, as we discussed in Chapter 2, then a new category of 'accepted' offspring could easily take account of ovum donation, if the rules were framed accordingly.

One question that needs some legal attention is what happens if something goes wrong during pregnancy. What are the obligations of the recipient woman and her partner? Equally, as with AID, the parental rights and responsibilities of the ovum donor need to be clearly nullified at some point in this process. Also the eventual disclosure of the donor's identity must be considered. As with AID, it is generally desirable. But it could be complicated if the donor has herself failed to have children. This would be a little different from the situation of an AID donor, who had never had children, because presumably he would not have been trying for them when he made his donation.

All this assumes so far that the donor has herself experienced involuntary childlessness and has been involved with ivf. This may not be so. She might offer to provide an egg by laparoscopy, for example, prior to sterilisation. This is already done for research purposes, as was the case with John Rock's patients.[23] Or a woman who would otherwise have no reason to come into hospital could also offer to be a donor. This is rather different. An analogy in this case might be brothers or sisters who offer a kidney to a sib. This is a much more considerable risk than laparoscopy, but it does happen from time to time, and in general this degree of altruism is thought admirable. If the analogy is apt, then we could allow female relatives or close friends to act as ovum donors. The genetic and psychological situation would be like sperm donation from a known near relative, which is often considered very problematic. It is obviously something that requires a lot of forethought.

Research continues on ovum freezing. This is still at the experimental stage, but will be feasible in the near future.[24] Some women donating oocytes for research might prefer that these were eventually fertilised and implanted, after storage in an ovum bank. This opens up the prospect of the commercial sale of ova, which might be marketed as having originated from particularly desirable women. This is as repellent as making available sperm from Nobel prizewinners. From a genetic point of view it is as misguided.

Embryo donation

Earlier in the chapter I said that the transfer of several embryos improves the success rate. For this reason women tend to be made to superovulate and several oocytes are removed from them. All are likely to be fertilised, but only a maximum of three transferred. This can leave 'surplus' embryos. For them the possibilities are destruction, use in research, donation, and, depending on the resources and expertise of the ivf team, freezing. The first seems something of a waste, the second does not appeal to everyone. In fact embryo donation has already occurred.[25] In effect it is prenatal adoption. I cannot see that it is significantly different morally or psychologically from ovum donation. It reminds me of the CECOS idea of a gift from one couple to another, or, I would add, to a single woman. The same kind of legal questions about legitimacy, responsibility and disclosure arise as with sperm and ovum donation.

However, there is a difficult issue here, which turns on how far embryo donation (or acceptance) actually is like adoption. In the

76

former case a cluster of cells is made available before a pregnancy; in the latter a child is offered to would-be adoptive parents during one and not transferred until after birth. Moreover, with adoption the original parents have a chance to change their minds. We take seriously their feelings about their child, even if they have not seen it. Donating an embryo ought to be different, since embryos are not children, and at the moment of commitment there has been no pregnancy. But what would be done if the genetic parents changed their minds during the pregnancy? It is all very well to say that is no longer 'their' embryo, since they gave it away, but how would we make that principle stick?

The ability to transfer embryos, and thus to donate them, has created a new category of living things, which are neither people nor property. We might be tempted to regard them as property stored temporarily at the hospital, as one might keep one's jewels in the bank or one's dog in kennels. But modern law regards the ownership of human bodies or organs or tissues as unacceptable.[26] At the same time pre-implantation embryos are not offspring either. They have not achieved personhood, so they cannot be relatives. They occupy an intermediate and indeterminate position, as potential offspring and quasi-property. People's rights and obligations as potential parents or quasi-owners are thus unclear. This ambiguity requires us to develop new forms of symbolic action, like the form of the CECOS donation or consultation over the use of transplant organs, to allow the different parties involved to express their interests and resolve their feelings.

Whatever legal or cultural device we use to indicate the commitment to donate an embryo, it has also to be subtle enough to take account of those cases, which ivf also makes possible, where an agreement is made for a surrogate mother to have an embryo transferred to her uterus and to carry it through pregnancy, only to return it to its genetic parents after birth. These complications are pursued further in the next chapter.

Embryo freezing

The issues are complicated by another technological option, which is embryo freezing.[27] I do not find it surprising that sperm survives refrigeration to −196 C, although many years passed between the development of a technology for producing such low temperatures and discovering how to freeze sperm without loss of function. But I am impressed by the robustness of living cells and embryos that allows them to be taken down to temperatures at which all metabolism has completely stopped, kept there perhaps for years, and then thawed. This has been done with the embryos of agricultural animals for some years. There are now a few babies, in Australia, in Holland, and very recently in Britain that have been born, having been frozen for several months as embryos.[28]

There are technical reasons for contemplating this. The minor surgery of laparoscopy is slightly traumatic and it seems that 2 days afterwards is not the best time for a woman's body to receive an embryo. Implantation may be more successful if several cycles go by, but that is only possible if the development of the embryo can be arrested by freezing. Embryo freezing could become a part of the routine procedure of ivf.

But consider the following example. A married couple are admitted to an ivf programme and five ova are removed and fertilised. They all develop. Three are implanted and the couple ask that the remaining two embryos are frozen. Nine months later female twins are born, and the couple ask that the embryos are kept in the nitrogen for the time being, whilst they decide whether they really want more than two children. They pay a fee for this service. They are then killed in a plane crash. A very similar case actually occurred in Australia in 1984, and as far as I can discover it has yet to be resolved.[29]

What now should happen to embryos such as these? Must they be destroyed if the genetic parents have given no indication of their wishes in such circumstances? Or should they be re-implanted in another woman? What account should be taken of the views of former parents' relatives, such as potential grandparents. In particular, what do we do about inheritance? After all the embryos could be potential heirs, and other people could have an interest in seeing they were allowed to develop, or not as the case may be. This particular problem can be solved by legislation that would allow people to inherit only if they had already been born at the death of the legator. In cases like this the ivf team must be allowed to decide what shall happen, and preferably offer them to a couple on their programme or to use them for research. They should consider the wishes of other people, who may have a view, but be under no special obligation to do so.

This is the position of the Warnock committee. They proposed that the storage of embryos be reviewed once every five years. Parents would not have the right to indefinite storage without discussing the matter with the hospital. In this way it might become clear that the original parents were very unlikely to ask for implantation in the future, and with their agreement, donation or research could be arranged. The report also suggests that if parents die within such a five-year period then it would fall automatically to the hospital to decide how the embryos should be used.

Or consider a slightly less tragic possibility, where the man is killed but the woman is not, and she seeks to have the frozen embryos re-implanted. This is analogous to the case of the frozen sperm considered in Chapter 2 and should be treated in the same way. Equally, suppose that it was the man who survived, who then marries again, but to a woman who is unfortunately also infertile. Rather than endure the stress of ivf they ask that the frozen embryos be re-implanted, and say that they are both happy with the genetics of the arrangement. It might seem odd but it would not be so different psychologically from the situation where a step-mother brings up the children of a recently deceased spouse.

Finally, suppose that neither party is killed but instead they get divorced. On re-marriage the wife requests implantation of the remaining two embryos; let us say that her new husband is himself infertile but they wish to bring up children, with which she will have a genetic connection. However, the first husband, who is consulted in the matter, objects. This is somewhat like a custody decision. I would have thought most people would support the woman's request here, since she experienced the stress of oocyte recovery and is now acting constructively, as it were, rather than obstructively. Of course, if the man himself has again married an infertile partner, then there really is a problem. One each?

Embryo research

Within the last three years research with human embryos has come to assume an enormous symbolic importance in the public imagination. The first reliable reports of human in vitro fertilisation appeared in 1969 and these were widely publicised. The scientists involved gave every indication that they planned to carry on with this work and to create more human embryos in vitro, as part of the research necessary to develop ivf as a medical procedure. But the idea of growing them in culture media did not strike most people as particularly controversial until almost fifteen years later. The present concern is undoubtedly part of a continuing debate over the morality of abortion and arises

79

also because of public scepticism about the value of research as opposed to therapy.

At the moment there are no regulations in the UK that apply specifically to embryos. Their status in law has still to be clarified. Use of foetuses and foetal material obtained from abortions and miscarriages is regulated by a code of practice drawn up in 1972.[30] With ivf researchers now voluntarily submit their research proposals to a committee of scientists and non-scientists set up by the MRC. A limit of 14 days on the development of any embryo used in research is required and the re-implantation of any embryo which has been the subject of a research procedure forbidden. Whilst administratively this exercise seems to work, despite some initial grumbling from research workers about what they saw as science fiction in the Warnock report, it will not be sufficient to stem the groundswell of public concern.[31] To do that some kind of legislation will be necessary. In this area especially it would be myopic politically and morally lax to rely on traditional acceptance of scientists' assurances that research is a self-evident good and the more of it the better.

What is the case for research? This comes essentially in two parts: one that relates to infertility investigations, and to ivf and its improvement as a procedure, and the other to investigations intended to contribute generally to embryology as a medical science. Firstly, it could be very useful to create embryos from a particular couple by ivf for diagnostic purposes, in order to try to find out why fertilisation is not occurring, or why implantation is not successful. Secondly, many aspects of ivf as a procedure could almost certainly be improved. Embryos grow more slowly in culture than they do within the body, and changes in the medium in which they are immersed almost certainly could help. Embryo freezing has a low success rate, and it ought to be possible to do it better. Then it would be very useful to discover how to get oocytes to mature outside the ovary. This would mean that they would not need to be removed so close to ovulation, which presently requires constant monitoring of the woman and creates a great deal of stress. Testing our understanding of maturation entails the creation of embryos which will not be implanted.

It would also be very useful to know much more about which sperm defects inhibit fertilisation and to test the safety of using micro-injection procedures that allow the introduction of single sperm into the egg. Then there is the question of spontaneous miscarriage, which affects around 100,000 women per year in the UK. This may arise because of chromosomal defects in the embryo, or because of some failure in the formation of a placenta from the pre-implantation embryo. Understanding more of how the extra-embryonic membranes differentiate from the original cluster of dividing cells could be very valuable medically. Again ivf could be used to diagnose specific problems in individuals or couples seeking assistance. This kind of research is very far from being disinterested curiosity. Its payoff for the alleviation of suffering could be considerable.

Less closely related to ivf is work on the identification of genetic defects. This I consider in Chapter 8. There is also the question of exploring what happens when organs or tissues fail to form properly in early development, and cause major medical problems, like spina bifida later on. Research into new forms of contraception, such as vaccines that inhibit fertilisation or implantation, is very important, as is the safety of any method to be tried out clinically. Research with fetal material is also of value in cancer research, for example in understanding how cells are self-regulated. Embryonic development involves a great deal of switching on of biological systems that facilitate growth and differentiation. Cancer also involves tissue growth, from cells that have gone out of control. Understanding normality in embryology can tell us something about the pathology in the formation of tumours.

Not all of this work is of the same priority. Some of it would have an immediate payoff, some would only form part of a very large picture in the medium term. Some of it could be described very precisely now, some of it is a little more speculative. But it would be silly to pretend that research based on in vitro fertilisation is trivial, and easily dispensed with. The more obvious lines of justification relate to ivf as part of the alleviation of infertility but work with more general application in contraception, medical genetics and oncology is also worth backing, if the case for resources is fully debated publicly. Scientists should expect to have to state and debate the case for resources and not to reply on public deference.

It is one thing to state that research would be desirable, another to argue that it is morally permissible. Here views diverge. There are basically three points of view. The first is that every human life begins at conception. Embryos have a right to implantation and continued development. Research with them is simply out of the question, and other procedures like freezing are also unacceptable. The second is that embryos are human entities and that we diminish ourselves if we treat them instrumentally as objects for research. The third is that human embryos belong to the human species but are not persons.

Their moral status changes as they develop. What we may do to them depends on what characteristics they have come to possess. I simply cannot accept the first of these. I outlined my objections to arguments of this kind in Chapters 3 and 4. The second argument is more persuasive, but again I cannot accept it, principally because I believe the symbolism of respect is not inconsistent with research, if the work is well thought-out. We do diminish ourselves if we use embryos thoughtlessly, or if we are indifferent to the feelings of those who make them available to research workers. But this is an argument for not using *any* embryos for trivial purposes; it is not an argument against using human embryos for research.

The third argument legitimates research. The question is how far it may be taken. That is to say, what moral principle will serve to identify the developmental threshold, after which research becomes unacceptable? At the moment research is only possible with pre-implantation human embryos. Implantation normally occurs about a week after conception and is complete a week later. To date the longest period for which a human embryo has been allowed to grow in a culture dish is 9 days. If development outside the human body is to go further, then some means would have to be found to simulate implantation, so that a placenta could form. We are thus discussing the acceptability of experiments that cannot presently be done. This is why it is very desirable to state the moral principle that should apply, lest we seem to be changing our standards to fit changing technical environments.

The relevant moral consideration is, in my view, the ability to feel pain and to form mental impressions. Until the embryo becomes a sentient being, able to experience distress, we can contemplate experimentation on it. The problem is to decide just what this means. One solution is to say the formation of neural tissue in the most rudimentary form represents this threshold. This is the basis for the limit of 14 days proposed in the Warnock report. But this is an extremely conservative interpretation of the embryology. For, since no electrical activity is observed in these cells at this point, nor is there anything like a nervous system or a brain at this point, it is implausible to suppose that there can be any mental activity. Another possibility is to argue by analogy from the definition of death now in use, which is based on the cessation of electrical activity in the brain-stem, and say that a limit should be set at the moment in development when such impulses can be detected in human embryos. This is not until some eight to ten weeks after conception. Almost certainly the capacity to form impressions vaguely similar to those associated wth conscious-

82

ness does not develop until very late in pregnancy, so setting a limit at this point would be both technically and morally supportable.[32]

Chapter 6

Surrogacy

The really curious thing about surrogacy is that within the last two years it has suddenly claimed our attention. It is not as if it has only recently become feasible, as the result of relentless technical change. Surrogacy can be very simple. In one form or another it has been employed for thousands of years. In marriages where the wife was infertile, the husband had a child by another woman, and at birth or soon after, the child was handed over to the infertile couple, to be raised by them as if both were its genetic parents. The woman who conceives and bears the child, and who is in this arrangement one of its genetic parents, is called the 'surrogate' or substitute mother and the arrangement itself surrogacy.[1] This simple terminology is not subtle enough now to cope with all the possible arrangements. In that sense technical innovation has changed something.

Virtually all of the recently publicised cases have relied on intercourse or artificial insemination to bring about conception. Only in a very few cases, which have got very little public attention, has there been any more novel technical activity. As far as I am aware ivf has not yet been involved to date, despite their frequent association in the media. However, ivf does introduce new possibilities which we must consider. So it seems that for once we are actually debating an issue ahead of time. This would be reassuring, were it not for the fact that we could have begun in earnest fifteen years ago.

If you itemise every new technological option within surrogacy you soon reach a dizzying number of possibilities and potential problems.

This is not so much tedious as bewildering. It is easy to lose any sense of how to think about the similarities and differences. It is best to proceed step by step, by considering first the simpler forms of surrogacy (intercourse or artificial insemination, followed by the transfer of a baby) before moving on to embryo transfer and ivf.

Informal 'domestic' arrangements

Surrogacy is discussed twice at some length in the Book of Genesis. Abram was advised by his infertile wife Sarai to have intercourse with her servant, Hagar, an Egyptian slave. This he did and Hagar found herself pregnant and began to mock Sarai. Abram told Sarai to deal with this situation herself and she drove Hagar from her household. However, an angel told her to go back, which she did and a son, Ishmael, was eventually born. God then made a covenant with Abram that his tribe would prosper and promised him that Sarai would soon conceive depite her age. As a token of the covenant Abram's tribe adopted male circumcision as a practice and Abraham and Sara changed their names. After the birth of their son Isaac, Sara began to object to Ishmael's presence in the household and he and Hagar were again expelled, much to her distress. The Lord again intervened and Ishmael eventually became a powerful man in his own right, as Abraham had been promised. The Genesis account also makes it clear that Abraham had concubines and that he had another wife, Cetura, by whom he also had children. The other story involving surrogacy concerns Abraham's grandson, Jacob, and it is even more complicated.

These stories describe the rise to power of the polygamous Hebraic tribes. They are told without embarrassment. For one thing the course that events are taking is discussed with God, by two of the people involved. He explains what is going to happen next and how everything can be solved, in a kind of endorsement of Abraham's tribal leadership and of Sara's use of a slave, who would continue to be part of the household. Hagar is not condemned; indeed her suffering is alleviated by the angel's appearance.

This, then, is a very special narrative description of how a specific kind of tribal society worked. However, there are some resemblances to one contemporary form of surrogacy, where a sister, close relative or friend bears a child for an infertile woman. Whether intercourse or artificial insemination takes place is not particularly important.

Recourse to the latter could minimise the sexual overtones. In such situations the surrogate probably does not disappear from the scene, and she acts out of a desire to help two people that she knows reasonably well. It may well be expected and understood that she play no role in bringing up the child, and ceases to regard it as her own.

We certainly know that this has been done from time to time but there is no way of knowing how common this kind of arrangement is, nor, as far as I can discover, are there any psychological studies of people in this situation. This is a pity because it has certainly been discussed as something that might be acceptable and as preferable to more commercial arrangements between strangers.[2] Again this is in effect a form of ovum donation, where the donor also carries the child during pregnancy and goes through the pleasure and pain of delivery. But this does constitute an important difference from pre-conceptional ovum donation. The equivalence is genetic only; practically, psychologically and morally, we are discussing something different.

It can also be considered a form of adoption, where all the parents are on hand to explain why the child was 'given away', if they so choose. It is a form of adoption where the original mother is not a distant and possibly irrelevant figure, and is thus like those cases today where the original mother has frequent access to the adoptive family. Also it resembles the situation where a man has a sexual relationship with two related women or close friends, one of whom becomes pregnant, although the other is more involved with the child-care and comes to be regarded as the 'social mother'. Assuming considerable maturity from all concerned, things could work out quite satisfactorily and people would grow in mutual respect. It is probably also the case that the man would have much the easiest role to fulfil here. Abraham seems to have told the women to sort things out for themselves, and his concern was for the son, Ishmael, that he was to lose. Typical, many women would say. The problem is to know what significance to grant these analogies. They appeal to our sense that people have coped, sometimes very well, under these unusual conditions. They mitigate the novelty, as we ask ourselves whether this *could* work. My answer, with respect to informal surrogacy, is that it could. But would it be likely to do so, and should we regard it as morally acceptable?

For example, we have been making optimistic assumptions about the course of pregnancy, the health of the baby, the delivery and the 'transfer' of the child, and the evolving feelings and responses of the major participants and other people. It is not hard to imagine the sort of resentment or jealousy or guilt that might emerge. But the same is true of many other ways of conceiving and caring for children. That is the point of raising the various analogies with step-parenting, 'blended families', where the children of several relationships are raised together, and non-monogamous households.

We must also consider motivations. Firstly, there are those of the people I shall call the receiving couple, since I do not think the alternative term 'commissioning couple' is apposite for these non-commercial, 'domestic' arrangements. For the sake of simplicity I am also going to assume initially that we are talking about a heterosexual couple, since that is the most likely situation, but clearly single men or women, or homosexual couples of either sex, could act in this way. Secondly, there are the motivations of the woman acting as a surrogate. There is a tendency to see her as someone acting very much on her own, which is probably a mistake. She may well have to consider the feelings of her friends and relatives, even if the decision to act is hers alone.

The most obvious reason for considering this kind of arrangement is infertility or an inability to go through the stress of pregnancy and labour for medical reasons. There might also be genetic reasons why the woman would not wish to have her own children, if she was not prepared to consider prenatal diagnosis, or if it was unavailable for the relevant condition. We must assume also that adoption is unaccept-able or not possible. Given that there are now many more people seeking to become adoptive parents than there are young, non-handicapped children to be adopted, scarcity alone might be the reason. The receiving couple may be thought unsuitable to act as adoptive parents, for 'trivial' reasons, such as their being slightly above an arbitrary age limit, or for serious reasons, such as their recurrent involvement with child abuse. And it is not unknown for the man to be particularly keen to have a child of which he at least is a genetic parent.[3] Or the woman may be reluctant to become pregnant or to go through labour for career or cosmetic reasons, and will have a lot of help with child-care. Obviously reasons can differ very considerably, some being factors which would at least evoke many people's sympathy, others being very worrying.

As far as the surrogate is concerned, the most obvious motive is altruism towards a friend or relative, but there are other possibilities, such as a desire to experience or re-experience pregnancy and childbirth or to become intimately involved with a particular couple, and to enjoy their gratitude or feel satisfaction in this 'ability' to bear children, thought of as a kind of skill. There is some information on what women seeking to act as commercial surrogates have felt, which I consider later on. We tend to think of them as rather different from those who act without payment, but this may not be so. Dr Philip Parker, a psychiatrist working in Michigan and involved with a surrogate agency, has published a brief account of interviews with 125 white women, applying to act as surrogates.[4] Of these 56 per cent were married, 20 per cent were divorced and 24 per cent never married; 60 per cent were working or had a working spouse; 89 per cent said they required a fee for their services, although one of Parker's claims is that the financial rewards were not the only motivation. 91 per cent had a previous pregnancy, and 81 per cent had had a previous live birth. Put another way, and not the way that Parker puts it, almost 10 per cent had not even experienced pregnancy before. Nor does he say how many still had their own children.

Some of those who had been pregnant, and he does not say how many, had found pregnancy, 'the best time of their lives'. 26 per cent had had an abortion and 9 per cent had relinquished a child for adoption. Some realised that they were still trying to come to terms with these losses. Most expected that they would experience some sense of loss, although some women felt that this would be outweighed by the pleasure of having 'given' someone a baby. These women are a self-selected group, applying to act as surrogates, and subject presumably to further selection. Nonetheless, complexity of motivation, revealed by Parker's study, seems likely to apply also in the case of women acting voluntarily.

These are the probable motivations. What problems might send things awry? The most likely psychological problems are jealousy and unrecognised attraction, the difficulty of giving away a child and of maintaining the commitment to do so, and maintaining the commitment to accept it, various forms of later regret, and problems of adjustment and disclosure for the child. We can immediately add to this the medical problems that could arise if the pregnancy is not straightforward or the baby is born with serious handicap and the possible legal problems of transferring parental rights and defining responsibilities if things do go wrong. That is already a formidable list, even though many also apply in adoption, which many people regard as a successful and desirable practice, certainly preferable to institutional life for children.

Amongst the possible responses to all this, two are very straightforward. One is to say that any form of surrogacy is deeply and profoundly wrong, because it violates what ought to be the exclusiveness of marriage, because it exploits a totally misguided sense of generosity, and because it will lead to untold confusion in the child and eventual regret from all concerned. That view has the virtue of simplicity, but it gives no guidance if we discover that despite strong condemnation people are still prepared to act without payment as surrogate mothers. This foursquare attitude derives from rather simple doctrine and ignores the psychological subtlety and generosity that make adoptive and multi-parental families work. Society as a whole has to be more imaginative.

Another response is that anything goes, that people's procreative and sexual liberties are paramount, and that they have the right of privacy to order their lives as they see fit. This, too, I do not accept, principally because the dividing line between personal responsibility and hedonism is very fine. In this case, as in many others, children's feelings are at stake. Perhaps adults can take care of themselves, but children deserve the most mature consideration of which we individually and collectively are capable. Much family law is practised on the basis of the best interests of the child and this should be no exception. At the same time, I argued in Chapter 2 for the acceptability of some non-traditional forms of parenthood, and shall take the same position here. In my view the question is not so much whether some marital and procreative arrangements are intrinsically unacceptable – and some are – but how they get set up.

Between blanket condemnation and blanket approval, there are two quite distinct intermediate positions. Both appeared in the debates about surrogacy in the House of Commons in 1985, occasioned by recent highly publicised commercial activity.[5] The first sees a ban on commercial surrogacy as an absolutely vital, immediate goal, but regards informal arrangements as beyond the reach of the law, as a family secret with which doctors and social workers should not meddle. Commerce is the main vice, and private actions, if kept within the family, are rightly and necessarily private.

The other point of view is also opposed to commercial surrogacy, though less militantly, but sees discussion, counselling and professional advice as highly desirable and far from meddlesome. Both positions endorse non-commercial surrogacy, cautiously but unmistakably. My own position is the latter.

In the Parliamentary debates in May 1985 about the proposed Surrogacy Arrangements Bill one concern of MPs who took this view was that the legislation was being rapidly prepared and would prevent doctors and lawyers from playing any role in non-commercial cases, for fear of the consequences. There was also some concern that people who might otherwise have sought advice would make their own imperfect arrangements instead. It seems that the terms of the Surrogacy Arrangements Act, now on the statute book, do not prevent lawyers from offering advice on the state of the law and its implications.[6] They may even see the receiving couple and the surrogate all together and charge a fee for their services. What they may not do is negotiate between the two parties for a fee. They may not arrange a contract for their own commercial gain. Also, as the legislation stands at the time of writing, it is not illegal for the surrogate to charge a fee. What she may not do is advertise, nor may anyone on her behalf. Surrogacy itself, with or without payment has not been criminalised.

The British government now assumes that the rapid growth of commercial surrogacy has been checked in the UK, and the status quo secured. A moral contagion being spread by American free enterprise has been stopped. At the same time in July 1985 the British Medical Association at its annual meeting in Plymouth passed by a narrow majority a resolution supporting surrogacy on a very restricted basis.[7] Some doctors feel that their medical and psychological advice could usefully complement legal guidance in a very few cases. Quite how this brave statement translates into practice remains to be seen. It is a straw in the wind.

The mix of expectation, self-denial and indebtedness involved in surrogacy must be potentially explosive. But it is not so different from other arrangements that one could say instantly that it would necessarily fail. For a few people it will continue to seem the most appealing option open to them. Some of them will make it work. Even so it is quite reasonable to argue that we still ought to try to stop it. For example, no doubt some incestuous relationships work, but we still feel that we have to try to prevent them, so profoundly damaging do they usually seem to be. Informal surrogacy is not so obviously a hazard and trying to stop it completely would fail, since it is relatively easy to conceal the fact of an informal arrangement.[8] Our objective must be to try to increase the chances that it will work out in as many different circumstances and anticipating as many problems as possible.

I have in my mind here the analogy of adoption. Many of the psychological issues that have to be considered by people applying to be considered as adoptive parents and by women or couples giving up their children for adoption are very similar. The selection procedure for the former is partly intended to make people more aware of their feelings about marriage, about infertility or their inability to have their own children, about other people's attitudes to them, and, above all else, about the needs of the children. Some people find all this a great trial, and it constitutes in their eyes one of the great problems with adoption. They resent the fact that their fitness to parent is on trial, by people in whom their trust is low. But if you believe that the interests of the children are paramount, then anything that can increase people's self-awareness is valuable. However, the difference between adoption and surrogacy is that in the former case people have little choice nowadays but to accept this procedure with all its frustrations, whilst in the latter they are autonomous, and not subject to selection and counselling. This may be precisely why they are contemplating surrogacy. One important step is to ensure there are no legal obstacles to the provision of advice to people who seek it. Thus doctors, lawyers, social service departments and charities should be free to deal with people who have come to them to discuss surrogacy. This would only cater for a self-selected group of people with some confidence and insight anyway, but I believe it is still worth doing. A further advantage of this approach is that it would allow the formalities of the adoption procedure to be used. The receiving couple could adopt the child, of which the father would be a genetic parent in any case.

Commercial surrogacy

So far I have been assuming that the surrogate mother is not paid. That this will be the case in all domestic arrangements is somewhat implausible. It need not be payment in cash, but instead new clothes, an exotic holiday, or a new car might be pressed on her by people anxious somehow to show their gratitude. Thus we cross an uncertain dividing line between a form of surrogacy that has received hesitant, heavily qualified approval, and another form that recently provoked the most forthright condemnation. What are for some people the most degraded and the most exalted states of womanhood have been joined in one phrase to damn it, as 'the prostitution of motherhood'.[9] Money apparently makes a world of difference.

Such is its power in all our lives that we fear the influence of money in sensitive areas where people need to be very clear about their motives. Nonetheless, this revulsion from finance is somewhat culture-bound. Manifestly it worries Americans less than Europeans, as the different attitudes to commercial surrogacy on each side of the Atlantic shows. What is it about payment for the service of surrogacy that troubles people?[10] When we talk about banning commercial surrogacy are we objecting to the idea of payment itself, or to the size of the payments, or is it commercialisation and a fear of what happens when informal trading becomes a business?

Let us consider a case in Britain that reached the courts in 1978, which illustrates the problems that can befall arrangements between strangers acting without any advice or control, where money changes hands. In 1976 a couple living together, with one of her children, approached a prostitute at Bow Street Magistrates' Court with an offer of £3,500 if she would bear a child for them.[11] This was refused but for £500 she offered to find someone who would act as a surrogate mother. This turned out to be a young woman of 19, who was set up in a flat and artificial insemination arranged. When the baby was born she refused to hand him over as agreed. More money was offered, and then a house, but still she refused.

The commissioning couple then took legal advice and the man sought access to but not custody of the child, no money having been paid to the mother. In his judgment on the case in 1978 Mr Justice Comyn roundly condemned the commissioning couple, and stated that the original agreement had no status in law, and ruled that no one should disclose any of these details to the child without permission of the court. At the same time he granted access to the father for 2 hours a week. This judgment was taken to appeal by the mother and the Court of Appeal decided unanimously not to allow any access at all.

This case epitomises many of the things that people fear are all too likely to happen. The commissioning couple strike a dubious bargain with someone who changes her mind. They contest her claim to the child. She then has to go to court, facing the possibility that the child will be taken into care. It is a traumatic way to be launched into parenthood. Nevertheless, we can only discuss this case because it went wrong. The logic of the situation is, then, that only our worst fears will be confirmed.

From the point of view of the commissioning couple the problem with surrogacy is its uncertainty. Some people are so desperate to have a child that they are prepared to pay a substantial sum to have the services of a surrogate specifically selected for her constancy of purpose. It is just that kind of consumer that commercial agencies seek to assist, someone with a lot more money than the average punter with which to buy a little more security.

According to one estimate commissioning couples must be prepared to part with around $31,000.[12] This is made up as follows: $12,000 to the surrogate, $3,000 on confirmation of pregnancy and $9,000 when the baby is handed over; a further $7,000 goes on medical expenses, $2,500 for psychological screening and counselling throughout the process; $3,000 is a living allowance during pregnancy; $1,000 for life insurance and $6,000 legal fees. On this basis roughly half goes to the surrogate, and a sizable chunk to the medical and psychiatric subcontractors to the agency. Whilst one agency claims that fees are negotiated between the commissioning couple and the surrogate herself and so may be somewhat variable, in other cases the couple are shown a catalogue of women, each of whom has a different price.[13] The case of Mrs Cotton, which caused such alarm in the UK in 1985, is broadly similar. She was paid £6,500, with a similar amount being paid to the American agency involved.[14]

Surrogates are recruited by advertising and it is said that significant numbers of women continue to apply. One agency director has claimed that 1,000 women answered his initial notice.[15] Some agencies claim that they only use married surrogates, but no one has been explicit about how they make their selection. They and the commissioning couple are likely to sign a document that lists all kinds of eventualities.

The whole exercise of psychiatric assessment, medical checks and legal consultations, is geared to one primary objective, the handing over of a baby, without any short-term complications. One man who runs an agency in Kentucky has claimed proudly, 'I haven't had one surrogate yet who has failed either to conceive or to hand over the child.'[16] All the money is there for that purpose to keep the surrogate mother firm in her resolve to hand over the baby. Much of the counselling and screening is done to see whether she will crack under the strain. How unlike adoption this is, where women can change their minds, and the procedures deliberately prepare everyone for this outside possibility. What the surrogate has been encouraged to sell is her autonomy; she has contracted out of being a free agent, because someone offered her a lot of money to do so. Fulfilling this role requires women to work at a set of feelings about that pregnancy and to deny a meaning that it would normally have. That denial, in which others may have to share, has to endure, first for nine months, and then for longer. The whole experience of pregnancy must be consciously shaped in a particular direction, away from any sense of permanent attachment or continuity.

I do not want to romanticise the implied attachment, and imply that women are necessarily having to struggle against basic feelings that they simply cannot still. For some it is like that, and for some not. But the most frequently quoted comment from surrogates is that they have constantly to work at their feelings, which are an ever-present threat to the enterprise. I cannot see this willed alienation as in any way desirable. The logic of the operation requires women to stick to an initial commitment, rather than work out fully what they really feel about their pregnancy, the developing foetus and themselves. It renders yet another area of our lives into mere work.

Proponents of the industry might say that the selection ensures that only those who can handle these risks are recruited and counselling helps them to deal with emergent problems. Yet there is no real guarantee that these assurances are worth anything. Surely those running the agencies will pursue their own material business interests, of maximum turnover, low overheads, reasonably satisfied customers and an expendable workforce, held in thrall by complicated contracts and a superficial concern for their mental state, while selecting and counselling and rationing psychological support? This, then, is my basic point, that I cannot see how we could ever have any meaningful guarantees that the very profound psychological effects at work in surrogacy would be taken seriously. Business is business and this one in particular needs people with suitably dulled sensibilities.

If it was the case that paid surrogacy would continue despite legal prohibition, with recurrent contests over children in the courts from bargains struck on street corners, then perhaps this more professional and expensive activity would be tolerable as the lesser evil. But it is not that the agencies clean up an industry, they create it. Unlike prostitution and the production of pornography, both of which are based on the culturally sanctioned exploitation of women, I see no evidence that commercial surrogacy will be sustained by well-established consumer demand. There is no case for surrogate agencies as a reform movement.

I do not object to the idea of a woman releasing a child of which she is a genetic parent and which she has borne during pregnancy to another couple, or another person. We allow that to happen with adoption, but to my mind adoption procedures set the standards. Nor do I object to the genetic aspects of the surrogate transaction. Genetic relations are not absolutes. People can give away their gametes – sperm or ova – and others can bring up children successfully of which they are not genetic parents. I see no moral or psychological obstacle here. However, the decisive question is the way in which and the time at which it is done. Sperm and oocyte donation is practically and psychologically different from handing over a baby, even if at the level of genetics the two situations are the same. To deny that is to ignore the experience of pregnancy, and I defy anyone to say that it is irrelevant.[17]

Thirdly, I also believe that the experience of going through pregnancy knowing that the child will be given up at the end of it can be handled. We expect that to happen with adoption and nowadays take some care that the feelings generated during this period are worked through. We also respect the autonomy of most women in this situation, bearing in mind also the best interests of the child. It is this element that is missing from commercial surrogacy, where a great deal of money and the resources of a business organisation are devoted to keeping a woman to the terms of an agreement, which has a very dubious status in law. Dealing adequately with all the possible problems that could arise here is beyond the resources and beyond the interests of any organisation that is being run commercially.

The onslaught on the agencies in the UK, that led to legislation in 1985, appears to show that they can be quickly shut down. But is there more that we should do? Legislation that resolved the legal un-certainty about the status of contracts between surrogates and those commissioning babies, as proposed in the Warnock report, would be helpful.[18] If there is just a possibility that the mother's claims might be overruled because she had entered into such an agreement, then people might be tempted to use them.

Another possibility would be to make payment itself illegal. At the moment in the UK surrogate mothers can be paid, although adoption proceedings cannot then take place if the fact of payment can be proven. One concern at present is that the adoption proceedings might come first, and payment be arranged discreetly afterwards. The foreknowledge of it would be in everyone's minds and thus influence the whole arrangement. It would also place the surrogate in a more vulnerable position of waiting for money that she was not supposed to receive. This cannot be desirable. Even if we see surrogacy as something we want to prevent, we do not want to create the conditions where the women we are trying to help are actually placed in a worse predicament, if they go ahead. Our goal should be blocking callous entrepreneurship rather than punishing surrogates.

The new technologies

Technological innovations complicate the debate still further. One of these is like Walter Heape's procedure, or its modern version which is used in the breeding of dairy cattle. It has several names, including embryo lavage, surrogate embryo transfer, and embryo flushing.

Consider the case where a woman could bear a child, and her husband or partner is fertile, but her ovaries produce no eggs, although her body produces the right hormones through the ovulatory cycle. No eggs could be removed from her because none are being produced. However, another woman is found, whose cycle is roughly in synchrony with hers. She agrees to act as an ovum donor, and at ovulation is inseminated with the sperm of the infertile woman's husband. Sperm from a donor could also be used.

The donor's Fallopian tubes are flushed out with saline solution several days later, in the hope of recovering a pre-implantation embryo. If an embryo is found it is examined and is re-implanted in the uterus of the infertile woman, in the hope that she will then become pregnant. The full version of this procedure was attempted in 1983 and, after a number of failures, a child was born in January 1984.[19] The donor then acts as a surrogate mother for just a few days if her egg has been fertilised. There is a possibility that the embryo will not in fact be flushed from her body, in which case the pregnancy will continue and some difficult decisions have to be taken.

In the experimental programme run by Dr J. E. Buster and his colleagues in California to test this idea, 12 infertile married women and their husbands took part.[20] 9 donors participated for some months, and they were asked not to have intercourse for the 5 days prior to ovulation and not take oral contraceptive pills. From 29 flushing procedures, 12 embryos were recovered. 2 of the 12 embryo transfers led to a continuing pregnancy, and one to an ectopic pregnancy in the recipient which was terminated. In addition one woman from whom an embryo was not recovered had a spontaneous abortion 9 days after her first missed period. One could describe the success rate as 7 per cent (2 pregnancies for 29 inseminations) or 16.7 per cent (2 for 12 embryo transfers). Donors and recipients gave their informed consent and the whole experiment was cleared by the ethics committee of the hospital. As far as I can discover the team have not published any more results since March 1984.

Proponents of the technique claim that it is safer for the recipient than ivf would be, were that possible for her, because she requires no general anaesthesia. However, there is a risk to the donor, of infection produced by the procedure itself, and there is also a risk that she will in fact become pregnant, if the embryo is not flushed out. As a procedure it is helpful to have a pool of donors, none of whom use contraception, who can be selected for insemination if their cycles happen to coincide with that of a recipient. There must be some stress of anticipation for donors, whether or not insemination occurs or not. In this case the donors agreed that they would terminate unintended pregnancies arising from the experiment. Presumably for this reason their cervical mucus was checked for the presence of sperm just before the artificial insemination, to see if in fact they could already be pregnant, having had intercourse with their partner.

There are some obvious potential problems here. For example, can you actually reassure donors by telling them that you check whether they might already be pregnant, since that check is also a comment on their veracity? Or if a particular donor does become pregnant after the flushing procedure and terminates the pregnancy, what do you tell the intended recipient and her husband? Do you tell individual donors when embryo transfer has led to a successful pregnancy? If not, do you tell them as a group the overall results of the experiment? Are recipients told anything about the donors' identities?

For me the most important question is the risk that the donor runs, from medical procedures with which she has no reason otherwise to become involved. This situation is therefore not comparable with ovum donation from women about to be sterilised. But suppose a sister or close friend offers to act as a donor. We have already considered voluntary surrogacy earlier in the chapter. This situation is comparable, both from the genetic point of view and psychologically. Where it differs is who is intended to become pregnant. This raises an awkward complication. For what if the donor is not prepared to maintain the pregnancy, if the embryo is not flushed out and implants? If it is a sister or close friend, the frustrated recipient will know what is happening. She could then ask her to remain pregnant. If she does not do, then the donor may feel guilty about 'succumbing' to an abortion. Psychologically and morally this seems very problematic.

Embryo flushing could be used in another way. I have no evidence that it has yet been tried. In this case the woman does produce mature oocytes, but is unable to continue the pregnancy, for any of a number of reasons, medical or otherwise. She and her partner have intercourse and several days later her tubes are flushed and the embryo re-implanted in a surrogate mother. But unlike the cases of surrogacy considered earlier in the chapter, and unlike the embryo transfers just considered, both members of the commissioning couple would be genetic parents of the child. That fact alone would make it appealing to some people. The commissioning woman would take the risks of the flushing procedure, and she might have to face a termination of her pregnancy. The surrogate would take the risks of pregnancy and have to part with the child. The problems with this are basically those of surrogacy generally, that the woman who goes through pregnancy does not retain, or is not intended to retain, the child. A rather similar possibility could arise through in vitro fertilisation. An egg is removed from a woman with blocked Fallopian tubes and fertilised with her husband's sperm, but is re-implanted in a surrogate.

Surrogacy in this more developed form has divided the roles of genetic parent, pregnant woman and social parent, as we considered in Chapter 1. This is not the kind of thing that people could do for themselves. It could arise if an ivf clinic and a surrogacy agency joined forces, perhaps offering package deals.

There are two basic questions here. The first is whether it is acceptable to transfer to a woman an embryo of which she was not a genetic parent, which she would carry through pregnancy, but which she would surrender after birth. This is what is sometimes called 'full' surrogacy, rather than 'partial surrogacy', where the ovum comes from the surrogate. The second related question is what the parental rights of the surrogate would be in this case and correspondingly what the rights of the commissioning couple would be. This question is raised by the obvious possibility of the surrogate mother declining to give up the child.

The 'fullness' of the surrogacy really makes no difference to the acceptability of the procedure. It is pregnancy which makes it a significant act, not the degree of genetic involvement. This requires that we grant primacy to the carrying mother. Her rights could then be transferred to the intended social parents after birth, by adoption. This would certainly allow for the possibility of wishing to retain the child and would offer a procedure for resolving that situation if it arose, by considering the best interests of the child and the desires of the various adults involved. The genetic parents – who in this case are also the commissioning couple – would be treated in effect as sperm and ovum donors, and have no parental rights to the child during the pregnancy, but would only receive them through the adoption procedure. This may seem rather odd, but I think that the nature of surrogacy requires it. By making use of a procedure that requires a woman to carry a child for them, the commissioning couple are asked to give up the rights they would normally have as genetic parents. They should be asked to trust the woman who is helping them have a child by conferring parental rights upon her, in the hope that they will recover them through the adoption process.

Chapter 7

Sex predetermination

People often say that it is only natural to want to select the sex of one's children, so easily do we describe our habits that way. By now we should know better. But how often do you hear future parents' preferences discussed and a biography for the child rehearsed during pregnancy? How often do we say, when told of a birth, 'Is it a boy or a girl?' These are the common place remarks of this and many other cultures. They are surprisingly hard to give up, even when you want to, so significant is gender to cultural identity.

But for some people the matter goes beyond vague preference and anticipation. So strongly do they want a child of a particular sex, or so pressing is the medical need to influence the genetic roulette, that they try to do something about it. What people have tried to do, what they will be able to do, its effects and possible responses are the subject of this chapter. So rapid are scientific developments in this area that they must be considered. Technical change alters the problems, although some of the basic issues have remained the same for years, if not for millennia. As elsewhere we have to sort out medical from other reasons. It turns out that the medical rationale for sex predetermination as such may soon disappear. However, the technology will either still exist, or be developed for other reasons, such as application in animal agriculture.

Sex predetermination could become very easy very quickly. It need not be 100 per cent reliable to attract widespread interests and it need not be widely practised to be significant. The mere fact that it was

100

possible would alter our ideas about conception. It would become something to decide not to do. The actions of even a few seeking sexual inequality would speak louder than the words of the rest of us, professing principled indifference as to gender.

The technical possibilities have provoked some wild speculations over the years, some writers phantasising extravagantly about massive swings in the sex ratio, leading to a chaotic descend into sexual deviation, even more prostitution, banditry and widespread disorder born of sexual frustration.[1] This kind of analysis assumes that mere numbers could be the prime mover of society. Yet there are communities with a preponderance of males or females, which are very orderly – monasteries and convents among them. So proportions alone cannot tell the whole story. Other, feminist writers have written of a huge threat to the continuation of the female sex, as the femicidal urges of husbands and fathers drive them to a 'final solution'.[2] It is hard to take such projections literally. But as allegories about deep-seated male distrust of women, manifest in many ways, not only a frequent preference for sons, it would be stupid to ignore them.

The effects of reliable sex predetermination would be more insidious and gradual, although there are countries like India and China, where son-preference is strong, and where a simple technology could have massive effects. In Western countries the prospect is not of a sudden lurch towards an overwhelmingly male population, but of continuing sabotage of women's self-esteem, as they are chosen as younger sisters. That possibility is deplorable, but the question of what to do about deep-seated preferences, which are often described as only natural, is difficult to resolve.[3]

The technologies

All cells in the human body with nuclei have 23 pairs of chromosomes, except for germ cells (eggs and sperm), which come to have 23 unpaired chromosomes. In other words germ cells contain half-sets of hereditary characteristics which come together in a specific combination (an individual ovum with an individual sperm) at fertilisation. In each set one chromosome is sex-specific. In human beings we call such chromosomes either X or Y chromosomes, because of their shape. Ova only contain an X chromosome, whereas sperm may contain an X or a Y. In the case of sperm all the chromosomes are tightly packed into the sperm head. In the case of mature ova prior to fertilisation, the chromosomes have still to separate themselves completely into two sub-sets. Fertilisation triggers the final jettisoning of one of them. But the eventual outcome will be that the fertilised ovum either has an X and a Y chromosome, and has the potential to develop into a male human being, or it will have an X and an X chromosome, and become a female. In both cases each of the 6 billion cells in the adult will have the same chromosome complement, of 22 pairs of autosomes and a pair of sex chromosomes. As always there are complications to this simple picture.[4]

We might, then, say that it is the sex chromosome from the sperm that determines the outcome. That is true in the human species at least, but is only part of the story. The process by which one sperm gets to fertilise an ovum has many levels of determination, many of them still unknown to us. Even though particular sperm-ovum encounters can only go one way, we have little idea what factors operate to determine which encounters will be set up, and which go further. Sex predetermination is about trying to influence a multi-stage process, of which we only understand one step. But clearly if we can influence which kind of sperm are involved in fertilisation, then we can determine the sex of offspring. Another strategy is to leave fertilisation to go as it may, and intervene at some point afterwards.

There are several possibilities to consider. The first involves procedures that try to influence the outcome of conception, through the timing of intercourse relative to ovulation or separating out one kind of sperm from another. Whether this is actually possible yet is debatable. The second involves the sexing of embryos prior to implantation. The third depends upon the establishment of foetal sex during pregnancy, by one of a range of methods, followed by the selective termination of pregnancy, and the fourth possibility is infanticide.

The last of these does not have a place here, but the fact that it goes on indicates attitudes to sex and gender, with a profound implication over how new technology is used. It continues today in countries where resources are extremely limited, or where very stringent population control is practised and distinct preferences exist for the sex of offspring. There have been a number of reports in recent years of the killing of female babies in the People's Republic of China.[5] In some peasant societies pressures on families and individuals may be very great. If the industrialised countries maintain others in a state of underdevelopment and enforced poverty, then it is somewhat hypocritical of us here simply to deplore how some people seek to cope, without also attacking the reasons for that poverty. Infanticide by them is a symptom and desperation the cause. Their actions are guided by a traditional view of women as a financial liability. The worrying thing about antenatal diagnosis is that it makes possible the kind of selection by sex once practised through infanticide. It undoes the effect of cultural change, at least for those with the money to pay the doctor.

Advice about how to influence the sexual outcome of pregnancy is ancient.[6] So too is folklore about its prediction. One very noticeable feature of all these ideas is the symbolism of left and right, which is itself a code for inferiority and superiority, left-handedness going with femaleness and the right hand with maleness. For example, one vintage theory held that sperm from the right testicle produced boys. The scrotum could then be bound up, or pinched during intercourse, if one was particularly keen to have a son. I find it hard to believe that men inflicted these procedures on themselves. Even less do I believe that aristocrats in the eighteenth century with estates entailed to male heirs actually resorted to partial castration on the same principle, particularly given that anaesthesia was only introduced in the following century. Or perhaps I, as a commoner, simply do not understand what *noblesse oblige* means?

Then there were theories in antiquity based on a supposed division of the uterus into two parts or chambers, as may occur in some animals. Women were required to lie on their right side during intercourse if a boy was desired, so that the foetus would develop on that side, and on their left for a girl. The same handedness appears in the traditional beliefs about establishing sex during pregnancy. Male foetuses were believed to lie on the right side, or to cause the right breast to swell with milk sooner, or the right nipple to become darker and so on. It is hard to know whether to regard such notions as pure symbolism, with no predictive value at all, or whether over the centuries they came to express partial truths. For example, Egyptian papyri describe pregnancy testing by soaking wheat and barley in urine, germination of the former indicating a boy, and of barley a girl. When repeated in the twentieth century the method seemed fairly reliable. It is known that some hormones that are found in the urine can influence germination. It would be silly, then, to dismiss all such notions out of hand.

In the late nineteenth century there were theories of alternating ovulation from left and right ovaries, so that couples were advised to try for a child of a given sex by counting cycles from a previous birth.

In present-day Western cultures this left-right symbolism seems to have disappeared. However, two other ancient ideas have been translated into a contemporary form; one is dietary control, and the other is the timing of intercourse relative to ovulation. Both of these enjoy a certain vogue.

One theory based on timing says that Y-bearing sperm move more rapidly through the mucus on the cervix and the uterine secretions, but that they survive for less time. If intercourse or insemination precedes ovulation, then an X-bearing sperm is more likely to fertilise the egg, since the others will not survive the wait for the egg. Intercourse at or shortly after ovulation favours Y-bearing sperm because they get there first. This idea was first advanced by Kleegman in the 1960s and taken up and widely popularised by Shettles in the 1970s.[7] Revealingly, an exactly opposite argument has been put forward by Guerrero and James, and they too have published data which purport to show success using their method.[8] Every so often someone tries to test these theories, the most recent case being reported in 1985.[9] This ran for 4 years in Australia, and involved 73 women, 21 of whom wanted a son, and 52 a daughter. For three months they kept a chart of their basal body temperature, which is known to rise slightly after ovulation, to learn when they ovulated. They were then asked to follow Shettles' method, in its less elaborate form. Of those wanting sons 7 were successful, but 14 gave birth to daughters: of those wanting daughters, 23 succeeded, but 25 had sons instead. One interpretation of the results is that those wanting sons would have done better not to enter the experiment and those wanting daughters did no better than they would have done if they had had sex when they felt like it.

Timing might teach one something about the human body and how to sense its cycles, but it seems useless as a method of sex predetermination. At the same time, the evidence suggests that Orthodox Jewish communities have consistently produced more boys than girls, often for centuries. This is believed to be because of the ritual abstinence from sexual relations during certain days in the menstrual cycle. In such communities the sex ratio approaches that in the general population when strict observance of Talmudic teaching is relaxed.[10] This is difficult to reconcile with the evidence from the most recent studies I have just mentioned, so we are left with rather contradictory evidence. Seemingly the timing of intercourse relative to ovulation can influence the sex ratio; but we have no idea as to why or how. It seems unlikely anyway that it could ever be a particularly reliable method. Perhaps one day a douche or gel will be put on the market and sold over the counter that really does make some difference, or, a diaphragm that would only allow one kind of sperm to pass through has also been discussed. Since sperm are only about 1/10,000th of an inch across, and the differences in size very slight, this idea is hard to credit.

Another possibility is vaccination. Mammals have a defence system that recognises substances and micro-organisms as 'not-self', and tries to destroy them. Materials of this kind are called 'antigens'. In the case of bacteria, such as those which cause disease, the antigens are molecules on the bacterial cell surface, which are recognised by the body as a sign of trouble. If the body has met them before, it goes into action against them much more vigorously. Vaccines are then weak versions of particular antigens, that prime the body against future threats. Men can make antibodies against their own sperm, which drastically lowers their own fertility. If there are sex-specific antigens on sperm, some distinguishing feature of X-bearing or Y-bearing sperm that the immune system can recognise, then vaccination for sex preselection would be possible. Another possibility might be for women to be vaccinated against sex-specific embryo antigens, so that embryos of that sex do not implant. Anti-fertility vaccines are a form of injectable contraceptive. One such agent, Depo Provera, which acts chemically, not immunologically, has already proved very controversial, because of enduring doubts about its safety, frequent medical indifference to its side-effects, and the injection of women without their consent.[11] All these difficulties would come up with sex-preselection vaccines, combined with problems of irreversibility, and I cannot see that the 'benefits' are either worth the risky experimentation their development would entail, or outweigh the problems in actual use.

All these technologies leave intercourse more or less untouched. If the sex of future children is thought really important then people will put up with some modification to the act of procreation – AID and ivf demonstrate their tolerance. One possibility is to process sperm, by filtration, or spinning in a centrifuge, or utilising the slight differences of electrical charge on X-bearing and Y-bearing sperm. The separated sperm, assumed to be almost 100 per cent X or Y, could then be used in artificial insemination. The separation is not something people could do for themselves in the bathroom, although they might take it home to use later. Indeed one procedure is being franchised by its developer to interested doctors in the US. The prospect of profit from commercial applications in agriculture makes people very reluctant to describe their work openly.

There are a number of papers claiming success, though the results are open to question.[12] One problem is checking the homogeneity of the sperm. Until recently the procedures only concentrated Y-bearing sperm, but in 1984 there was a report of a girl being born after using a method to separate X-bearing sperm, which is more useful in dealing with sex-linked disease.[13] One cannot conclude from this that the method works, but only that it does not destroy sperm. Nonetheless the consensus is that sperm separation will soon be possible in cattle and human beings.

Another possibility is to determine the sex of embryos, before or shortly after implantation. This was first done in 1975, by removing a small piece of tissue derived from the developing embryo.[14] In 1978 a cow embryo was sexed by removing cells from it at a very early stage and examining their chromosomes, and in 1982 calves were born from embryos which had been bisected, one half being sexed by chromosomal examination, and the other transferred. In 1983, according to a report in *Nature*, two female calves had been sexed as 6-day-old embryos, by identifying the presence of a particular molecule on their cell surface, having been flushed from the body of their mother before implantation.[15] Superovulation of high-yielding cows, embryo transfer and embryo freezing are now routine in the dairy industry. This technique, if it works, would allow breeders to select the much more valuable female embryos from the male, avoiding the expense of feeding a surrogate-mother cow whilst it carries the less valuable male calf. Six months later Robert Edwards was reported as saying that he found such work 'extremely interesting' and that he hoped to begin establishing the sex of pre-implantation human embryos, using DNA probes that recognise part of the Y-chromosome.[16] This would involve dividing embryos into two at around the 8-cell stage, examining the chromosomes in one half with the probes, and freezing the other half for re-implantation at a later cycle, if it turns out to be the right sex. This could be done with all the embryos available from a particular cycle. As far as I am aware, this work has still to begin.

This procedure could be used to prevent sex-linked genetic disease. This I consider in the next section, but two points stand out immediately. One is that the DNA probes here simply recognise a section of the Y-chromosome. They represent a relatively quick way of determining its presence, but they can also be made much more specific, so that they can identify individual genes on particular chromosomes and thus indicate whether an embryo has acquired the gene for a given genetic disease. In other words one can go further than simply testing for the sex of the embryo. The other point is that this is a specialised form of ivf, done not because of infertility, but in an attempt to avoid sex-linked disease. The success rate is likely to be lower than that of ivf in general, because of the embryo division. That places limits on the diffusion of the procedure at present, but it also suggests yet another way in which ivf could expand, and underlines why it is such an important technology.

Another way to establish sex is to allow implantation and fertilisation to take place and to determine the sex of the foetus at some point in pregnancy. There are a variety of ways, in which this can be attempted, of which three stand out. The first is the newest. It involves obtaining tissue from the trophoblast, or the tissues that surround the developing embryo after implantation and that grow into the lining of the uterus. These tissues derive from the original embryo and are therefore genetically identical to it. Such tissue can now be removed by biopsy after six to eight weeks, with some risk to the continuation of the pregnancy, and the results of DNA analysis of the material removed are available within 48 hours. This procedure is still at the experimental stage. Its advantage is that it allows the pregnancy to be terminated relatively easily within the first trimester.

Secondly foetal cells, obtained from the amniotic fluid around the foetus by a procedure known as amniocentesis, can be cultured. This can only be done at around the 16th week of pregnancy, and growing the cells and observing their chromosomes itself takes several weeks. Therefore, if abortion then has to be done, it is rather late and a more complicated and traumatic procedure. But selective termination of pregnancy after amniocentesis to establish the sex of the foetus has been done for some twenty years now.

The final possibility is ultrasound examination of the foetus. Whilst it is possible to generate images of the embryo quite soon after implantation using high frequency sound waves reflected from it, the establishment of foetal sex in this way has to wait until the external genitalia form and are visible. This usually means waiting until the third trimester of pregnancy, and even then it can be very difficult to catch a glimpse of the relevant organ. It is said to be easier if the foetus passes urine during the scan, when the angle at which it emerges is informative. This then is not a particularly reliable method, except perhaps in the hands of the most skilled ultrasound operators.[17]

In all, sex predetermination begs for three basic questions to be considered. The case for it on medical grounds, to deal with sex-linked disease; the extent to which people would try to use some form of sex predetermination and the potential after-effect; and, what should be done now to control the technology.

Sex-linked disease

In some genetic diseases the gene involved is carried on the X-chromosome. The condition is not only sex-linked but also X-linked. Haemophilia is a well-known example. Since the Y-chromosome carries just a handful of genes, males who inherit the gene on their X-chromosome will suffer from the disease concerned. Unlike females, who can also inherit the gene from a parent, they have no second X-chromosome that can compensate for the gene causing the problem. It follows that women can also pass on the gene to daughters, who may in turn pass it on, but female carriers of the gene will not suffer from the disease. Women can also inherit the gene from their father. This is necessarily so, if he has the disease. He may also be unaffected by the condition yet still pass on the gene. This would arise if there was a mutation in the gene on the X-chromosomes in his sperm. Furthermore, affected males cannot pass on the gene to their sons, since they must pass on a Y-chromosome to a son, but all the daughters of affected males will carry the gene.[18]

All this is just a shade complicated to take in abstractly. Another way of expressing it is to say that if the transmission of the gene is through the male line, the chance of male children being affected is zero, but the chance that female children will be carriers is 100 per cent. If the gene is inherited from the mother, the chance of males being affected is 50 per cent, and the chance of female children being carriers is 50 per cent, and the chance of their being affected is zero.

One of the classic examples of a recessive X-linked disease is haemophilia-A, where the body fails to make any of the blood protein clotting factors VIIIc, so that sufferers are constantly vulnerable to uncontrolled bleeding. Another is colour-blindness in its several forms. Another is Duchenne muscular dystrophy, where the muscles waste away in childhood. About 1 in 10,000 males suffer from haemophilia.[19] The pattern of inheritance was first recognised in 1803 in Philadelphia, although the genetic basis was only fully elucidated after the development of the chromosome theory of inheritance in the twentieth century. A classic illustration of it was produced by the geneticist J. B. S. Haldane, using the descendants of Queen Victoria in the English, Prussian, Russian, German and Spanish royal families.

Queen Victoria was herself a carrier of the haemophilia gene, probably having inherited from her father. Her eighth child, Prince Leopold, was haemophiliac, and died young, although not before having a daughter who was a carrier. Two of Victoria's daughters, the Princesses Alice and Beatrice, were also carriers. Both women passed the gene to some of their children, and thence it was transmitted to the next generation, amongst them Prince Alexis, the son of Csar Nicholas II. Increased medical knowledge, the lottery of inheritance and revolution have all contributed to the apparent disappearance of the gene in the contemporary branches of these families today.

However, it was only from the mid 1950s that the antenatal establishment of foetal sex could be used in counselling people who stood some risk of having a haemophiliac son. Foetuses found in mid-pregnancy to be male could be aborted, although 50 per cent of them would in fact be perfectly healthy. Over the same period the treatment of haemophilia has also improved, with the production of dried clotting factor concentrate extracted from large amounts of donated blood at considerable expense. However, this means that haemo-philiacs are vulnerable to viral diseases transmitted in blood, including hepatitis B and AIDS. Couples are now being advised not to have children and the families of haemophiliacs are experiencing great difficulties as they and their children are ostracised without any justification.[20] In 1979 an antenatal diagnostic test for haemophilia itself was developed, using foetal blood. In 1984 the clotting factor gene was successfully isolated and transferred to bacteria by genetic engineering, so that they synthesised minute amounts of clotting factor.[21] This offers the prospect of human Factor VIII from non-human sources. It will also allow DNA probes for the haemophilia gene to be used in antenatal diagnosis at an earlier stage in pregnancy.

When the gene is transmitted through the female line, AID is no help. Ovum donation might be, if the donor was known not to be a carrier, which could now be determined with DNA probes. Dealing with a particular family might reveal which sisters were not carriers. Conceivably this might encourage a very few people in such families to consider ovum donation. Most people, I believe would opt for some form of antenatal diagnosis instead. AID is a possible solution when disease is inherited through the male line, for example if a haemo-philiac man wants to have children. But of course the need for it is far less, since although all daughters must be carriers, all sons must be unaffected. Equally, in some cases like that of Queen Victoria's father, the Duke of Kent, the problem was undetected until his affected grandson was born, at which point it was Victoria herself who needed the advice, which was simply not available.

It might seem that sex predetermination offers a simpler solution. This is only partly true. Firstly, female carriers of the gene are likely to learn of their status only through the birth of an affected son, or of a possible problem through the birth of an affected son to a sister or cousin. Today even if we screened all women to locate those who carry the gene, which would be expensive and problematic, this would not locate cases occurring through mutation. Sex predetermination cannot help everyone or reduce the incidence of haemophilia to zero.

Secondly, for those who actually know that there is a risk for them and want to have another child, antenatal diagnosis is now possible. This means opting for an early abortion, in the 25 per cent of pregnancies where the foetus is male *and* has the haemophilia gene. However, some people will not be able to accept antenatal diagnosis on moral grounds. Their choices will be to have children anyway or to opt for some form of sex predeterminaton. Sperm separation is likely to be somewhat unreliable for a very long while, and, if people object to the selective termination of pregnancies they are quite likely to object also to the selective initiation of pregnancies through ivf. Even for those who do not have such moral objections, ivf followed by the genetic analysis of embryos before implantation might be appealing, but its success rate in establishing a pregnancy with selected embryos would probably be lower than with selective termination. Also, it is not obviously less stressful. The point is, then, that sex predetermination might help the avoidance of some sex-linked disease, but in rather fewer cases than might appear at first sight. There is now only a weak case for pursuing research in human sex predetermination, given the present availability of antenatal diagnosis, or the alternative of ovum donation.

Sex predetermination: would people use it?

Sex-linked disease is serious enough for those who confront it, but this is perhaps only 1 family in 1000. Questionnaire surveys indicate that far more people might contemplate sex predetermination simply because they want a child of a particular sex.

In 1980 two sociologists, A. Ramanamma and U. Bambawale, published an article on two hospitals in India, where, in the late 1970s, amniocentesis had been utilised for the selective abortion of female

110

foetuses.[22] They called their piece, 'The Mania for Sons', resolutely indicating the pervasive male chauvinism in Indian culture sanctified by Hindu tradition. In one case 92 out of 400 women consulting the hospital in 1976-7 came simply to find out the sex of the foetus, with the intention of terminating the pregnancy if it was female. These women also said that they would not request a termination if it was male, even if it had a hereditary abnormality. In the other hospital studied, out of 700 women attending the hospital in one year, 450 were informed that the foetus was female. Of these 430 were aborted. Of the 250 pregnancies in which a male was predicted, all proceeded to term. To Western readers, this is shocking.

This data is undeniably puzzling: for example, the sex ratio of 450 females to 250 males begs the question of whether the women had some familial predisposition to produce female children. Whilst they might find their way to this hospital because of that, it is rather unlikely that 700 women would do so in one year. Another possiblity is misdiagnosis. Amniocentesis involves removing amniotic fluid from the uterus, through a syringe. In the fluid there are foetal cells and maternal cells, which can be grown on a culture medium, and the chromosomes examined. In order to separate the two kinds of cells they must be grown for several weeks. However Ramanamma and Bambawale report, without commenting on it, that the results were available after two or three days. These are just the conditions under which one would expect maternal chromosomes to be mistaken for foetal chromosomes, and the sad conclusion we must draw is perhaps 20 per cent or more of the results were wrong. That would give us a sex ratio of about 1:1 again.

Whatever, the significant factor is that 95 per cent of the pregnancies where a female was predicted were terminated. One likely possibility is that these women already had several daughters and may therefore not have been so concerned about the survival of a female foetus, and were going to try to conceive again in the hope of having a son. The figures as they stand are ambiguous. They could indicate that people wanted at least one son, (and had been tested after the birth of several daughters) or they could show a general preference for male children from every pregnancy – (a mania for sons.) Either way it is clear that amniocentesis was being used here to abort some foetuses simply because they were believed to be female and the hospital knew this. The women came predominantly from upper- and middle-class households, people for whom another child could not have represented an impossible burden, although they might have been reluctant

111

to provide a dowry for a daughter, as custom required.

In 1984 Viola Roggencamp described the operation of a private amniocentesis clinic in Amritsar, run as a business by a husband and wife team, who charged considerable fees for their sex-typing service.[23] She interviewed one woman who had been pressured by her husband to go through this procedure, and who had obviously been upset by it. This clinic has been the focus of controversy in the Indian press and Parliament, but several people have suggested that the Indian state is simply not in a position to enforce a ban on such initiatives, outside the hospital system.[24] Clearly, even if people only have occasional recourse to this form of sex preselection, the technology is capable of occasioning considerable suffering for women, because of what their men demand. Several of the other technologies already discussed, such as sperm separation and chorionic biopsy, could also be taken up by medical entrepreneurs, operating from the margins of their profession.

Sex predetermination, if it works, confers control over birth order. Historically this is why sociologists have asked people what their preferences are and demographers have weighed the consequences of their being able to realise them.[25] For example, if people believe that their ideal family would be one with at least two sons, then they might well have them straight away, rather than wait for them to turn up in the genetic roulette. These people would then achieve something approximating to their ideal more rapidly than they would have done otherwise and families would tend to be smaller. Of course the assumption is that people have distinct sex-preferences, that they have a 'stopping rule' of some kind (four is the absolute limit, or we'll stop only when we get two sons, or whatever) and that they are willing and able to stick to it.

As such this approach 'fits' small, planned, Western families with low mortality better than larger families where the expectation is that some children will die. Indeed if you believe that you need many babies as insurance for old age, then sex selection is not particularly relevant, since the chances are high that you will get at least a few children of the preferred sex. Much depends on how desperate the situation of the growing family is and the traditions that dictate what work men and women do. If things get bad then girls may be neglected, unless they too are vital workers on the land. One situation where such 'rational' calculation might go on is in countries like the People's Republic of China, where very strict population control is now enforced, where a quota of one child per family exists and where there is traditional strong son-preference. These are said to be the reasons that have caused a resurgence of female infanticide in China in recent years. It was there in the early 1970s that the techniques of chorionic biopsy were developed, to make selective abortion easier.[26] In general it is very difficult to forecast the likely effects of sex predetermination in less developed countries, but even the prospect of a major onslaught on female children should be sufficient to make us abandon the vague hope of slowing population growth in this way.

Small planned families are now very common in Britain and the US, so what might happen here? Questions about sex preferences and family building are asked in fertility surveys. One such was analysed by Pebley and Westoff in the United States in the late 1970s.[27] Two findings stand out, that most people prefer a more or less 'balanced' or symmetrical family and that many men and women want a son as the first child. Thus in this survey, of those married women who had had no children, 45.5 per cent wanted a boy as their first child, 20.3 per cent a girl and 34.2 per cent would be happy with either. Of those women with one boy already 72 per cent hoped for a girl next and 15 per cent a boy. Of those with one girl 68.4 per cent wanted a boy next and 14 per cent a girl. This is the balancing effect expressed in preferences, and it stands out more clearly when women have two children already, both of the same sex. However, at the same time 10 to 15 per cent of women say that even then they would be happy with either, which may be because they do have no preference, or because they hope to have a large family.

One implication of figures like these, which seem to be fairly stable over time, is that if sex predetermination were to be practised on any scale the population sex ratio would not shift significantly. So much for anarchy born of sexual scarcity! Another implication is that there would be fewer families of all boys or all girls, which is not terribly interesting or worrying. But what is significant is the fact that fewer women would be first-born children.

There are problems in using such survey data. It is based on answers to hypothetical questions, from women interviewed on their own. The question assumes an understanding of how sex predetermination would be practised. Women sometimes think that some form of abortion is implied, which biasses their replies. The predictions also assume that the actuality of choice will not change anything, and that people will keep on making the same choices, even when their effects become clearer.

You can also ask more direct questions, about whether people would actually use the technology themselves. There are quite a number of such studies from the US, most based on what students think. A recent one canvassed the views of 236 undergraduates at 2 American universities.[28] 24 per cent of the women and 23 per cent of the men endorsed using the techniques, but 59 per cent of the women and 62 per cent of the men said they would not do so. That is vaguely reassuring, even though these are people who are not representative of American society as a whole and who are unlikely to have started a family. But of those who said they would be interested in sex predetermination 81 per cent of the women and 94 per cent of the men said they would prefer first-born sons. So although the proportion of people who say that they either have no preference or would not try to realise it seems to be rising, there is still a substantial minority who would seek to have sons first. The implication is that we can expect a subtle shift in social experience as more men grow up knowing that they had been chosen to be older brothers – the children you have first –and more women recognising that they were intended to be younger sisters – the children you have later, to balance the males. Maybe the shift is not so very subtle?

So what do we do?

The developments described here are disquieting, in two ways. Firstly they are all too close to realisation technically. Secondly, their utilisation is deeply rooted in what many people think of as legitimate desires and preferences. To the unreflective they are only natural, and the suffering and regret that they may cause are not even an issue. So what do we do, cogitate, legislate or agitate?

This largely depends on what we take to be the problem and what we want to prevent. We can ignore sex-predetermination as a preventive strategy for genetic disease. This problem can be tackled in more precise ways. We also need not excite ourselves with the prospect of a disorderly predominantly male (or female) population. A re-enactment of Klondike conditions without the gold is unlikely. The problem is the cumulative psychological effects of general preferences about sex, enacted family by family. One form of this is a passionate desire for a first-born son, or a male only child. Another, which is less worrying, is the desire to balance a small family, with at least one of each. The former suggests an ingrained male chauvinism, coloured by deep scepticism about what women could ever do, the latter some recognition that the sexes are different, and that both have their value, although males are conceived first. I predict that medical entrepreneurs will try to service both desires in the future, using improved technology, principally developed for use in agriculture. This would give a new meaning to the word 'spin-off'. I also predict that preferences, attitudes and actions will divide somewhat in the way that votes divide, with two opposed blocs of sentiment, one in favour of sex-predetermination, the other strongly against, separated by floating voters uncertain quite what they ought to do. The question, as in electoral politics, is how best to win over the middle ground.

One option might be an extension of equal opportunity legislation to cover conception. Just as we believe that men and women should have equal access to promotion and pay, so one could argue that people should have an equal opportunity to be born as male or female. Interfering with the apparent randomness behind the 'natural' sex ratio would be unjust. On the face of it this is rather attractive, reworking the legislation of the 1970s to fit the reproductive technologies of the 1980s. But there are problems. Firstly, the sex ratio at birth is not quite 1:1. Slightly more boys are born than girls, in most countries. Would legislation require us to adjust this figure to equality? Secondly, and more importantly, equal opportunity legislation at present applies to persons. But to whom would this new extension apply? More importantly still, it is somewhat unlikely that everyone would accept as reasonable or necessary the implied restraints on their reproductive behaviour.

Throughout this book I argue for reproductive freedoms, and the removal of some of the barriers that the conventional wisdom would otherwise impose. I am reluctant therefore to support the re-introduction of constraints, even though I disapprove of what people intend. Thus, in the case of Mr Graham's Germinal Repository I think the whole idea is very wrongheaded in several ways, but I do not believe in legislating against eugenic sperm banks. Gentle mockery and critique are more defensible and likely to be more productive in the long run. In the case of sex predetermination, too, it is better to retain a degree of procreative freedom, and to argue with people over how they use it, than to deny it to them.

My justification for extending the range of the permissible has been to enhance people's confidence and self-respect. Confident, well-informed people, with access to advice, rather than subject to professional coercion, think more deeply and act more consistently and with greater respect for others. People forced to live according to very restrictive conventions come to resent the fact and wriggle around the rules, often at a cost to themselves and to others. And confidence, although not a hereditary trait, is passed on, unto their children's children. But the offensiveness of sex-predetermination lies in its direct challenge to women's confidence in their sex. It enables men to ask them to collude with the reproduction of inequality in the next generation. Perhaps, then, because it strikes so directly at the values of a possible society, free from sexual inequality and self-doubt, we should allow ourselves to act more aggressively against it?

I think not. The kind of reproductive freedom I have in mind is not a haven that once created need never be left. It is something that has constantly to be hazarded, argued for and re-defined. It cannot be locked in place with legislation. It has to be re-created over and over again in the decisions of millions of people, some of whom will act in error. Attempting to corral people with legal restrictions is a step in the wrong direction. If, as I believe, it turns out to be very difficult to get such legislation on to the statute book, then the shouting and arguing will be more valuable and educative than the result, and the means should become the end. If it is very easy to legislate, then we should all beware, because others with less benign and generous intentions may step in ahead of us, before we realise it. With that thought, that reproductive law can menace as well as protect our basic freedoms, we turn to a field whose history bears that out, human genetics.

Chapter 8

Antenatal diagnosis:
an inglorious past, a hi-tech future

The idea of genetic disease

Like begets like, but how and within what limits? The science of genetics exists to answer such questions. What began as the study of seed shape and petal colour in sweet peas is now a discipline of immense scope, embracing the reproduction of birds, bees, bacteria and the human species. It is fundamental to modern biology. It will be fundamental to the medicine of the next century. In particular it will play an increasingly important role in determining our attitudes to health and disease, normality and abnormality, even valid and invalid existence. It is hard to exaggerate its likely cultural and psychological effects over the next ten to twenty years, as our technical ability to select the kinds of people who are born increases.

The first flowering of genetics came in the 1920s with the demonstration that hereditary traits, like white eye-colour in the fruit-fly, were specified on the chromosomes, the rod-like bodies in the cell-nucleus. The second period of growth began in the 1950s and continues apace today. It has taken our understanding of the mechanisms of inheritance to a finer level. Whereas the gene was originally an abstract notion, a determining factor transmissible between generations and responsible in an unknown way for the appearance of particular traits, we now know it is something quite concrete and manipulable. It forms part of a programme through which each individual organism is built up. Like is begat by like

117

because of the transmission of a structured instruction-set, a genetic programme, from one generation to the next.

The basic laws and concepts of genetics come from its first growth period earlier this century. Their explanatory success fuelled the enthusiasm of some scientists and doctors from both Left and Right to create a new science of population management called eugenics.[1] They believed that modern civilisation had arrested the process of natural selection, allowing the genetically unfit to survive, and, worse still, to reproduce without let or hindrance. This was evident, so British eugenists argued, in the rejection rate of British conscripts in the Boer War. Selection had become unnaturally weak and had to be re-asserted. For the general good a measure of reproductive quality control had to be imposed, either through the provision of contra-ception or through the involuntary sterilisation of those with no sense of eugenic duty to the race. This is negative eugenics. Some eugenists went further and claimed that those with superior genetic attributes had a duty to procreate, even to try to outbreed the less desirable. This is positive eugenics. The most organised version of this nonsense was the Lebensborn programme, involving members of the SS and German girls who matched some supposed Teutonic ideal in the 1930s. Herman Muller's plan for 'voluntary germinal choice' was a milder version, requiring civic self-denial in the name of genetic responsibility.

Such ideas sound barbarous today, but it is only sixty years since they were a commonplace, and negative eugenics an almost un-controversial part of professional middle-class opinion. Many of the pioneers of family planning, abortion law reform, artificial insemi-nation and social welfare provision were committed to the science and politics of eugenics to some degree. Habitually we tend to equate eugenics with the racial engineering of the Third Reich; but sterilis-ation laws were also passed in the US and Scandinavia in the 1920s and seriously debated in France, the UK and the Soviet Union. It was very far from being a fringe activity or the fetish of a particular culture, even though scientists' support for its more extreme manifestations lessened during the 1930s.

After the war knowledge of the Nazi policies and the demonstrable flaws in some of the early ideas further eroded its appeal. But eugenic sentiments and institutions still survive in a variety of milder forms. The Repository for Germinal Choice, inspired by Muller, is one; Lee Kuan Yew's exhortation to middle-class women of Singapore to reproduce for the good of the nation is another.[2] In a different sense (to be explored in the next three chapters), so too is much

118

humanitarian concern with genetic disease today, when it focusses less on helping people make decisions that feel right to them, and more on reducing the incidence of a particular condition. We can do much harm if we impose societal considerations on individual reproductive decisions.

The question as ever is how to select from the technological options those that will genuinely enhance people's freedom and give them the power to live their lives as they see fit. The constant temptation with modern medical genetics is not to re-create the racial hygiene of the 1920s, but to deprive people of some of their autonomy, in the belief that one acts for their own good. What lives on is perhaps not eugenics but a kind of genetic paternalism.

The idea of hereditary disease is not new. Some maladies, folk traditions have said for centuries, 'run in families'. In the nineteenth century physicians often spoke of a constitutional weakness or susceptibility, which they called a diathesis.[3] Vague and ambiguous though this notion now seems, it has the virtue of emphasising an intrinsic inability to adapt to particular environments. It is the interaction of an individual with his or her surroundings that leads to the manifestation of disease. It is both the range of possible adaptation and the range of possible environments that create the problem.[4]

The implications of this are two-fold. Firstly, it may be possible to change the environment. For example, the genetic disease phenyl-ketonuria (PKU) is caused by the lack of a liver enzyme that converts the common amino-acid phenylalanine to tyrosine. We take in phenylalanine with our food; individuals with PKU are unable to break it down, and high levels build up in their bloodstream and this usually leads to mental retardation. If they eat a special diet that lacks this amino-acid, the problem never arises.

Secondly, much depends on whether we feel the environment can or should be changed. For example, we tend not to call hereditary astigmatism of the eye lens a genetic disease, because by modifying the way in which light reaches the lens from the environment we can mitigate the problem. On the other hand, we tend not to devote the considerable resources required to make the physical environment less challenging for people with spina bifida, who are likely to suffer from a degree of spasticity. Every intrinsic deficiency is linked to an 'implicit environment' that sets limits on what we are prepared to change. It follows that what we classify as a disease is a convention, dependent partly on our technology.[5] The lower boundary of the category is particularly significant, for it defines the limit of tolerable variation at any given time. Beneath this level you have mere variation; above it, disorders that create disease. As our technological expertise changes, what we consider a severe problem requiring genetic intervention changes too, as does the kind of solution we may entertain. For instance although PKU can be treated fairly successfully, by dietary or environmental modification, some scientists are now contemplating gene therapy as a possible cure.[6]

Nowadays we attend more closely to genetic diseases because the background rate of infant mortality has fallen. They now represent a significant proportion of the illness that leads children into hospital, perhaps 30 per cent, and a greater percentage of deaths in childhood. Many more conditions are now known to have a genetic basis, 3,000 to 4,000 at least, although many of these are exceedingly rare. Few are treatable, which means that some of those affected die soon after birth or go into a fatal decline in late infancy or adolescence. In other cases the problem is serious physical handicap, needing constant medical attention. Something like 2 per cent of all live births are of children suffering from genetic disease or hereditary problem. In the UK that is around 16,000 children each year. If we say that an average large maternity hospital in the UK deals with about 3,000 births a year, that amounts to 60 affected children a year, or about 1 a week. This is not the same as the number of cases of severe handicap, which occurs for a variety of reasons, including brain damage during delivery.

A very considerable proportion of medical activity is devoted to advising and counselling rather than organising treatment. This is partly because there usually is no treatment, and partly because people want to know the risks if they have more children. In this sense it is rather different from other areas of medicine. It is also much more kin-oriented. The doctor's role is often to generate the information to allow people to take a decision. This is likely to require information from relatives, who in the process may discover medical facts about themselves that are unwelcome. The trauma spreads out to envelop a kin group, rather than just a household.

The complications do not end there. Firstly, the statistics of inheritance are often quite subtle. For example, it is quite possible for a significant proportion of individuals within a population to carry one copy of the gene for a particular condition (say 25 per cent), and to be perfectly healthy themselves, whilst only a small proportion of babies being born (say 2 per cent) inherit two copies of the gene, one from each parent, and are desperately ill as a result. Both the apparent disparity in numbers and the difference between the carrier state and that of having the disease has to be explained to the people concerned. Secondly, inheritance is probabilistic. Every germ-cell is created by sorting a set of genetic instructions into two sub-sets on a largely random basis. Fertilisation recreates a full set again, as the chromosomes from the sperm pair off with the set of chromosomes left in the egg. Whether one or two copies of any given gene have been passed on by this shuffling is always a matter of chance. We can often state the probability of a particular gene combination occuring. Hard though it is to convey and accept this roulette in the hereditary process, it is one of the facts of life in sexually reproducing organisms.

We have four kinds of genetic intervention to consider. The first is where couples with an affected child, or whose relatives have had one, seek medical advice to learn what has gone wrong, and to discover the risks that they run if they try to have another child. For some conditions that is all that can be done, even today. This form of genetic counselling is very much the preserve of medically qualified specialists in the UK, less so in the US.[7] Sometimes they overestimate their ability to convey helpful information to people in a state of shock. Others, with fewer qualifications but more time, can sometimes be more effective. It is not just empathy that is needed. Getting the diagnosis right is vital.[8]

The second possibility comes in when a pregnancy is under way. With some conditions the genetic status of the foetus can be determined during pregnancy and a decision taken as to whether it should continue. This is antenatal diagnosis. For some people the medical termination of pregnancy is unacceptable on moral grounds, and antenatal diagnosis can serve only to reassure the couple concerned or allow them to prepare themselves for difficulties ahead. Even without such fundamental objections, this procedure raises some awkward questions. These are considered both in this chapter, which deals with developments up to the present, and the next one, which looks to the near future.

The third possibility is to determine people's genetic status some while before they try to have children. This is genetic screening and its aim is to forewarn. Those who find that they are carriers of the gene for a particular disease can use this knowledge if they contemplate having children with another carrier. They may then end that relationship or decide not to have children or ask for antenatal diagnosis. Nowadays they may also contemplate recourse to artificial insemination or even ovum donation. Screening is problematic, because of the difficulty that people have in assimilating information of this kind, unless they are well counselled. The record to date is mixed, as we shall see in the last section of the next chapter.

Finally, there is gene therapy. This gets a chapter all on its own, partly because it is still not quite feasible in human beings, but mainly because it is such a serious issue, representing a significant step beyond what we do now. Gene therapy is on the verge of practicability. Within ten years it will be an established way of helping a few people with genetic disease. Our situation now is not unlike that concerning ivf in the late 1960s. The technology seems such an awesome prospect, but within a decade will be here to stay, for good or ill.

Antenatal diagnosis introduced

Attempts to obtain foetal cells were first made in the 1950s.[9] This was done by passing a needle through the abdominal wall into the fluid-filled amniotic sac to draw off some fluid, in which cells can be found. This procedure is known as amniocentesis and is much in use today. By the mid-1980s between 2 and 3 per cent of all continuing pregnancies in the UK were investigated in this way in mid-trimester.[10] This means that it is sufficiently uncommon for many women not to know anyone else who has been through the experience. It is not possible until 16 weeks, so that any resulting decisions have to be taken even later, well after foetal movements are likely to have begun. If the pregnancy is terminated, it involves the premature induction of labour, which can be emotionally difficult and is not without its medical complications.

For some people such action is morally quite unacceptable and criticism is increasing, particularly in the US, of abortion performed even when the foetus is known to have a serious genetic defect. They would also reject my choice of words in describing this situation. Their argument is that any abortion ends the existence of a human being, with a right to life, even though it is performed here before the foetus has become capable of independent life and become 'viable'. Nowadays abortion law in many countries holds that the State does not have the right to prevent a woman from ending her commitment to a pregnancy until the point of viability, at around 28 weeks. The anti-abortion groups oppose that philosophy. They argue on doctrinal grounds that any human life is sacrosanct, no matter how gruesome or brief the prospects after birth. This is a matter of dogma, which I cannot accept. Attitudes to abortion tend to be highly polarised. What

122

stands out from the opinion surveys is the high proportion of people, except those at one extreme, who regard the diagnosis of foetal abnormality as a valid reason for pregnancy termination. This distribution of views is also to be found in women who are themselves pregnant.[11]

Another, probably more widely shared argument is that our concern for the handicapped will diminish if we allow ourselves to terminate affected pregnancies. Interestingly the pressure groups and charities concerned with disability and handicap divide on this question. Some see prevention as the primary goal, which will not prejudice our care for those people who will continue to be born; others argue the contrary.

Very often the arguments involve someone saying, 'If amniocentesis had been available to my mother, I would not have been born; but even though I have to struggle, I am making something of my life, so I am glad that amniocentesis was not available to her.' As a piece of rhetoric against selective abortion this is usually very effective psychologically, because of the power in the affirmation that life is worth living. But as an argument against antenatal diagnosis, or not coming into existence, it will not do, because it is based on a comparison between existence and non-existence, which is meaningless. No one can possibly know what it would be like not to have existed. Any one of us might rejoice in or regret our existing, which is fair enough; but what we cannot do is to make comparisons between what has happened and did not. Similar considerations apply where people are tempted to sue a doctor or their parents for having allowed them to come into existence. However, if one could show that some sort of negligence occurred, and that one might not have had particular problems, then the case is different. Courts are therefore starting to look slightly more favourably on actions for 'wrongful life'.[12]

These are rather subtle distinctions, which tend to get lost in the rough and tumble of social life. For example, people sometimes say, 'I am very glad that I am alive, but I am worried that people think that people with my degree of disability should not have been born.' The crucial question is what happens to general social attitudes if some kind of eugenics is practised.

This argument about the erosion of social concern is very difficult to evaluate. An analogy with polio may help. Since the introduction of vaccination in the 1950s the incidence of polio has fallen dramatically in the developed countries. As cases have become less frequent, fewer people know what the disease is and why vaccination is worthwhile, so that the general level of immunity is now falling and the number of cases is starting to increase. Successful prevention programmes can shift public attitudes in undesirable ways. We want to keep in the public imagination both the seriousness of the disease and the need to care for those few people who still contract it, some of them through vaccination programmes themselves. But at the same time surely no one would argue that we should abandon vaccination, either to remind ourselves what polio is like or to stiffen our resolve to care for future cases? The same applies to genetic disease. Prevention can lead to indifference, but this is not a valid argument for not trying to prevent the birth of children with genetic disease. It is an argument for working to prevent indifference.

Two fundamental moral objections to the idea of antenatal diagnosis can then be set aside. But, as always, it remains important to ask how things work out in practice. The question is not just whether a particular procedure or policy is compatible with our basic values, but also whether the way in which medical assistance is organised recognises the hopes and fears that people have, fully recognises their autonomy and builds their confidence in what they choose to do. This is particularly important with genetic disease, where people's fears for their children are considerable, so that they need every support, but, medical mores being what they are, people and their offspring may be punished for allowing a particular pregnancy to continue. The quality of care is defined not by the number of 'problems' prevented, but the number of choices that people are offered and the confidence they are helped to develop in making that choice in an informed way.

Down's syndrome and neural tube defects

Antenatal diagnosis today is largely concentrated on the detection of Down's syndrome and the so-called neural tube defects. Down's syndrome used to be called Mongolism. We now honour the nineteenth-century physician, John Langdon Down, by using his name, but eschew his racist term for the condition. In 1959 Jerome Lejeune in France showed that it occurs in individuals who have three rather than two chromosomes No. 21 (trisomy 21). This is used as a diagnostic indication antenatally, although Lejeune is a longstanding opponent of abortion.[13] Four per cent of cases of Down's syndrome arise instead because of the translocation of a section of chromosome

21 to another chromosome, often one of the pair No. 14. The person is then born with what seems to be 46 normal chromosomes, but one of the 14th pair is a composite, with an extra bit from a No. 21 added on.[14]

The chromosomal abnormality, trisomy-21, seems to be associated with maternal age, although this is not the only cause. In some cases the condition arises because the father's germ cells contain an extra chromosome. Nonetheless one preventive strategy is to offer amniocentesis to women over the age of 35. Policy at the local level may differ. Some UK hospitals take 38 as the age limit, some 40; some wait for women to ask. The risk that a woman of 35 will have a child with Down's syndrome is about 0.25 per cent, or 1 in 400, rising to 5.7 per cent at age 45.[15]

Even so the majority of children with Down's syndrome (65 per cent) are born to mothers less than 35. This is because even though the risk is very much less, very many more women have children at younger ages. Indeed 94 per cent of women giving birth each year are under 35. Overall the birth rate of children with Down's syndrome in the UK would be around 1.2 per 1,000, or about 1,000 a year, without antenatal diagnosis. But a proportion are prevented by amniocentesis. This is probably about 6 per cent, which may seem a surprisingly low figure. The reasons include the limitation of screening to older women, some of these women not being told that antenatal diagnosis is available, or their deciding not to use it, or some of them deciding to have amniocentesis, but, usually in the face of medical criticism, allowing the pregnancy to continue.

As one might expect amniocentesis carries a certain risk, of injury to the foetus or of causing a miscarriage. Three large trials have been done in the US, the UK and Canada, which tried to match a large number of pregnancies with and without amniocentesis. There are problems in interpreting this data. One judgment of the risk of foetal loss is around 1 per cent.[16] This is slightly greater than the risk of having a Down's syndrome child until women get into their forties. The studies also show a slight increase in the number of babies with respiratory difficulties after amniocentesis. These are aggregate figures, and doctors vary in their degree of skill. Thus the actual risk at a particular hospital may be greater or less than that overall. Moreover, some women may be so concerned at the prospect of having a child with Down's syndrome that they are prepared to face a slightly greater risk to the pregnancy itself in order to be reassured.

Another major reason for antenatal diagnosis is to detect neural tube defects, principally spina bifida, where the column of tissue around the spine fails to close, and anencephaly, where a very substantial proportion of the brain is missing. In the former case the medical implications after birth are somewhat variable, ranging from mild disability to severe spasticity and mental retardation. In the latter case the baby is stillborn, although once the situation has been discovered it is usually less distressing to terminate the pregnancy. The rate of incidence varies from about 5 per 1,000 live births in Wales to just over 3 per 1,000 in East Anglia.

In 1972 it was discovered that high levels of a protein made in the foetal kidney, called α-foetal protein or AFP was very often a sign of a neural tube defect. Nowadays in most British health regions women attending antenatal clinics are offered a blood test at 4 months, to look for AFP, since some will cross the placenta and appear in their blood stream. If the level is high, a second, more reliable, test can be done using amniocentesis, to check the level of AFP in the amniotic fluid. If that is also high, and other indications, such as an ultrasound scan, confirm the diagnosis, then the pregnancy can be terminated. Again all the tests take time and are done in mid-trimester and they induce acute anxiety. In the medium term it is likely that tests will become more specific, quicker to perform and cheaper.

Results from the West of Scotland from 1976 to 1982 show that 140,471 women had maternal serum AFP tests. Of these 1,350 (1 per cent) went on to have amniocentesis. About 100 terminations a year took place as a result. The rate of anencephaly and spina bifida fell from 4.3 per 1,000 live births in 1976 to 1.7 per 1,000 in 1981.[17] There is thus a pyramid of involvement with the commoner forms of antenatal diagnosis. Large numbers of women are now offered the preliminary AFP test; that is to say, it is done as part of their antenatal care, without much explanation. A very much smaller number have further tests, of which the great majority produce negative results, so that only a very small proportion of those originally tested face the decision of whether to terminate their pregnancy.

The situation in the United States is somewhat different and interestingly so. AFP screening is not done routinely, although it is available. In 1979 the Food and Drug Administration was asked by various drug companies to approve test kits used to detect the protein. They envisaged a market worth $25 million. The American College of Obstetricians and Gynecologists (ACOG) and various interest groups, such as the Spina Bifida Association, opposed this because they felt that without adequate quality control in the application of the tests and adequate knowledge of how to interpret them from doctors, more harm than good would be done. What was in doubt was whether a disparate private medical system could be adequately regulated and good practice and advice guaranteed.[18] Approval was delayed until 1983. In 1985 the Department of Professional Liability of the ACOG circulated a letter to all its members warning them that if they did not ensure that they advised women that the test was available then they might be open to a charge of professional negligence, a warning that in the US context physicians would take very seriously.[19] However, the reservations about lack of expertise and counselling are still relevant today, so that the diffusion of the tests is driven by economic and legal pressures rather than by an ability to provide the best possible care.

It is now being suggested that low levels of AFP indicate Down's syndrome. If this were confirmed as a reliable test, then many more cases could be detected antenatally. At the moment it seems that it is not an indication that is sufficiently reliable to be of use.[20] The implications of some kind of general test for Down's syndrome eventually becoming available are very considerable, however. Firstly, a far larger number of women could have the option of antenatal diagnosis; secondly, far more resources would be needed to meet the demand, assuming there was an increase; thirdly, popular conceptions of Down's syndrome would change; and fourthly, more women would go through the cycle of investigation, waiting and relief or decision.

The experience of antenatal diagnosis

As a proportion of all abortions performed each year the number of those done for genetic reasons is extremely small, though it differs from them to some extent in that it ends a wanted and perhaps a planned pregnancy. As already mentioned, they must be done in mid-trimester, although very recent technical developments are now changing this, as we will consider in a while. But the delay before being able to act means that women may well conceal the social fact of pregnancy, or deny it to themselves, until they get a welcome result from the tests. In Barbara Katz Rothman's phrase it becomes a 'tentative pregnancy'.[21]

Various studies have been done of women's and men's experience of antenatal diagnosis, although there are still too few of them.[22] They show that, although most people endorse the procedure and its general

127

availability, quite often, despite their general moral position and religious commitments, a great deal of concern and emotion simply goes unrecognised. Moreover, some people would like to make use of amniocentesis, even though they know they would not have an abortion.

In 1978 Sarah Bundey sent a questionnaire to 325 women graduates, asking them how they felt about amniocentesis to detect Down's syndrome.[23] These were women who had been at Girton College, Cambridge, between 1953 and 1957 and were between 38 and 43 at the time of the survey; 268 replied. Of these 54 per cent had 3 or more children; 5 per cent had had or were intending to have children after the age of 38. Only 10 of them had already had amniocentesis. Thus only a small proportion had actually faced this experience; but, except for their educational background, they were like most women who face it. In total 222 (83 per cent) said they would have amniocentesis; 44 (17 per cent) said they would not. Only a fraction of each group volunteered their reasons for their decision. Those opting for amniocentesis tended to mention personal knowledge of how burdensome bringing up a handicapped child could be; those opting against tended to cite fundamental moral objections to abortion. Nine women said they would have amniocentesis, if the need arose, but that they would not necessarily terminate the pregnancy. Thus, even though a majority of women are likely to opt for amniocentesis, and are more likely to do so in countries where abortion has been legalised for longer, this view is not universal, and some women act with very mixed feelings, that will not be quickly resolved.

In 1981 Donnai and his colleagues discovered that most of a group of 12 women followed up after a 'genetic termination' within the previous four years had complicated feelings about it, even though all but one felt they had acted rightly.[24] What is remarkable about this paper is not so much what they found, but their admission that they had not expected it. They now try to ensure that women have follow-up visits from a health visitor to allow them to talk things through.

Some of the most detailed research on women's experience of antenatal diagnosis has been done by Wendy Farrant.[25] From her interview studies she reports recurrent failure by the health service to acknowledge the emotional complications of amniocentesis, to provide enough information about the purpose, nature and risks of the techniques involved, to structure the choices at each stage in an unbiassed way and to allocate enough time and money to counselling during and after the investigations. These are major shortcomings, that significantly reduce women's autonomy, despite the general overall endorsement of antenatal diagnosis. The point is not that women do not want the option of antenatal tests, but that they do not often like the terms on which the assistance is offered.

For example, of 112 women interviewed after amniocentesis, 28 were unaware that it is associated with a risk of miscarriage and 96 were unaware of other possible risks to the newborn infant. Of the 16 women who knew of these risks, 12 of them had obtained this information from sources other than medical staff, and the 4 who had been told by doctors had been given this information in an attempt to dissuade them from going through with amniocentesis, which they had themselves requested.

Another feature of current practice, clearly revealed by Farrant's research, is the frequent requirement that, before the amniocentesis is done, the woman agrees to a termination if an abnormality is diagnosed. One argument offered in justification is that amniocentesis without such a commitment is a waste of money. In other words a few people are denied the option of preparing themselves for the stress to come on grounds of cost. Even on its own terms this argument is very weak, since that very preparedness may save money later. Despite that this is very poor argument, since in effect it punishes people for their views, even though some may have over-estimated their ability to cope. What makes things worse is that the majority of these women would not have to consider a termination anyway, since the initial, less reliable, diagnosis may well not be confirmed by later tests. If they know that they are not even eligible for amniocentesis, they go through pregnancy with the anxious thought that something might be amiss.

There is a very important point here, which must be spelt out. It is a worthwhile objective to seek to reduce the numbers of children that are born with serious physical and mental handicap, stemming from genetic conditions and many other causes. Antenatal diagnosis has a role to play here, as does better education, better nutrition, more appropriate antenatal care and higher standards in obstetrics. But this aim becomes corrupt if we start to build degrees of coercion into it, supposedly for the general good. Our priority then must be to give people the time, the information, the assistance, and the freedom to decide for themselves what use they want to make of antenatal diagnosis, without presuming what their attitudes to abortion should be, one way or the other, or what they must do to qualify for it.

Very similar considerations apply to another preventive strategy for neural tube defects. There are now claims that these can be prevented by providing vitamin supplements in pregnancy, the implication being that even though a predisposition to have affected children may be genetic, the problem is made much more likely by a deficient diet. The problem is that the causes of neural tube defects are unknown. Family studies give only ambigious answers. suggesting both genetic and environmental causes. And there is no obvious biochemical reason why the absence of a particular vitamin should trigger this particular defect.

In the absence of more biological information, the only way to test the idea of using vitamins is to compare the numbers of cases that appear in the offspring of matched groups of women, some of whom are given extra vitamins in pregnancy, and some of whom are not. To complicate matters both the women concerned and their doctors would have not to know whether they were actually being given vitamins or not, since otherwise they might be tempted to give or take some vitamins, just in case they helped. This then would be a double-blind trial, like that sometimes used to test the effectiveness of a particular drug.

Obviously setting up a trial on this basis is extremely problematic. Reports in 1983 that the MRC planned to do so led to an outburst of criticism.[26] For example, it places the women concerned in a very difficult position psychologically, facing just a chance of nurturing a foetus with a neural tube defect and perhaps having an affected child, in order to help future generations, when there is already evidence that this can be prevented. Unfortunately, this evidence is insufficient for anyone to say that vitamins will help. Given the complexity of the uncertainties, how could one counsel someone to take part, if she had doubts? Since it is very likely that most women would have some anxieties about participating, it seems impossible to run such an experiment without some unintentional or well-meant coercion. Wendy Farrant's results above suggest the limited amount of counselling and explanation that women may receive. Carolyn Faulder's study of double-blind trials of different surgical strategies in dealing with breast cancer suggest the same thing.[27] Therefore I do not believe that the consent of most of the participants would be free and informed, given the time available to talk things through. Despite criticisms like these the experiment has gone ahead.[28] In the United States a different approach is being adopted. This will involve following birth records at two hospitals to see if mothers of women who give birth to affected children are less likely to have taken vitamins than women who have given birth to unaffected babies. The methodological problem here is to decide whether the comparison is valid, whether the two groups of women are in fact comparable.

There is more to say about the experience of dealing antenatally with the prospect of genetic disease, but to get to grips with it we need to consider a little more basic human biology.

Single-gene defects

At a global level the commonest genetic diseases are those which affect the production of red blood cells. These include sickle-cell anaemia and the various forms of thalassaemia. In both cases affected individuals inherit a genetic instruction set that prevents them from making the oxygen-carrying protein, haemoglobin, in its correct form. In the case of sickle-cell anaemia the alteration in the genetic programme is extremely specific. One chemical component of the haemoglobin molecule is mis-specified, with the result that when the protein is being synthesised the wrong chemical unit takes its place in the growing chain of such units. Consequently the red blood cells, which contain millions of such molecules, fail in their task of transporting oxygen, at particularly crucial moments. Then they tend to collapse into a sickle-shape and block up the blood vessels. Tissues are starved of vital oxygen and the body can only take so much of this trauma before its systems start to fail. This may be soon after birth or it may take longer, depending on the environment. With thalassaemia the set of genetic errors is larger, but again the common problem is that not enough functional haemoglobin is being made. Transfusing blood can do something to help but that brings its own problems in its train.

Both these single-gene defects are recessive. Chromosomes come in pairs. Each of the thousands of genes on any one chromosome has a partner on the corresponding chromosome. (As we saw in Chapter 6 that is not true of genes on the X-chromosome in males.) These gene partners can be identical or they can differ somewhat. It depends on which genes have been passed on from each parent. In effect each gene is a set of instructions in linear form, which specify the order of the amino-acid sub-units from which protein molecules are assembled within the cell. Like written text these sets of instructions can be corrupted. This almost always leads to problems, so that the molecule specified by that particular gene is built wrongly, or is only partly

131

assembled or the synthesis never even starts. If one copy of a recessive gene takes the correct form then the body will make enough of the substance, from the instructions in the unaltered gene, and no health problem will appear. One could say that the genetic defect recedes behind the functionality conferred by the normal gene, whereas with dominant conditions which appear even when only one copy of the relevant gene has been inherited, it dominates. Recessive genes may be passed on for a number of generations, without making their presence felt, whilst dominant genes lack this reticence.

We can now often determine whether an otherwise healthy individual carries one copy of a particular gene. Such people are sometimes referred to as 'carriers', or more technically as 'heterozygotes' or as heterozygous for a particular condition. Correspondingly people with two identical copies of a gene we call homozygotes. I am going to use the term carriers, even though I have always found it slightly alienating, because it reduces people to their genes. Speaking of 'heterozygotes' is even worse. It is, alas, a very convenient shorthand.

Single-gene defects are usually inherited in very simple ways. This allows very simple predictions of the risk of their occurrence to be made. For example, if two people are known to be carriers of a recessive gene, they have a 25 per cent chance of having a child which has inherited two copies of the gene, one coming from each parent. There is also a 50 per cent chance that each of the children will themselves be carriers. On the other hand, if someone is a carrier of a dominant gene for a hereditary condition, and necessarily also suffers from the condition, there is a 50 per cent chance that each child of this person will also be affected. Rather more conditions are recessive than dominant. One of the more common dominant conditions is Huntington's Chorea, which causes slow but unrelenting neurological failure and death in middle age. We consider this disease in the next chapter. Risks of 25 and 50 per cent are usually considered much too high to take, and this is why many people opt for antenatal diagnosis, if they know that they are carriers, and if it is technically possible.

One of the great problems with many hereditary conditions is their relative rarity, compared with infectious diseases. Even though the number of carriers of a particular gene may be quite high, the vast majority of them will never have any reason to know their genetic status. Very often diagnosis of carrier status is retrospective and comes about simply because of the birth of an affected child. This means that the incidence of that condition will not fall very far below the 'natural'

132

level, because only those who know themselves to be carriers will be in a position to do anything about it. There are other complications of rarity. Firstly, doctors may misdiagnose the problem, never having seen such a case before, and give the wrong advice. Secondly, some people find it impossible to believe that their child could be so sick when they themselves are so healthy. Men sometimes doubt that they are the genetic father. However, some diseases are much more common in specific racial or national groups and this allows community-wide programmes to be set up which alert carriers to their situation. Then, depending on how carriers use that information, the rate at which affected children are born can start to fall. Incidentally, note the eugenic undertone in the last sentence. I mentioned the incidence, not people's feelings.

There are about 200 million carriers of inherited haemoglobin disorders in the world, i.e. 5 per cent of the world population.[29] Between 200,000 and 300,000 affected children are born each year, most of them in less developed countries. This remarkable situation has arisen because individuals with one aberrant copy of the globin gene enjoy a slight degree of protection from malaria. All the hundreds of different mutations in the genes that control haemoglobin production represent evolutionary adaptations to the environmental challenge of malarial parasites. What we see now is the result of natural selection over millennia. The genetic price that has to be paid is that a small percentage of offspring who inherit two copies of a mutant gene will die in infancy. Thus in many tropical and sub-tropical countries, where malaria is or used to be endemic, these haemoglobin disorders are not uncommon. In small, geographically separate, countries like Cyprus as many as 25 per cent of people may be carriers. Even though in many non-industrial countries many children die in infancy, aspects of genetic disease may be recognised as distinctive. In West Africa the babies who die are called 'those who are condemned to return again and again', indicating that repeated deaths are to be expected in certain families.

The evolutionary geography of such blood disorders has been disrupted in recent centuries by forced and voluntary migration, from tropical Africa, from the Far East, and from the poorer Mediterranean countries to the richer states of Northern Europe and America. Thus in the UK sickle-cell anaemia is largely confined to the black population of Afro-Caribbean descent, and thalassaemia to people of Cypriot, Greek or Italian origin and to sections of the Asian community.[30] Their medical experiences are often complicated by various forms of racial discrimination, distrust of the health service, and moral, medical and cultural traditions at variance with those of the dominant culture.[31]

One of the first attempts to tackle such problems at the community level occurred in Greece in the late 1960s.[32] A group of scientists began to test people in a swampy region on the Aegean coast, and to tell them whether they were carriers of the sickle-cell gene, which can be determined from a simple blood test. They also tried to explain the genetic significance of this information, in the hope that the numbers of carriers marrying carriers would be reduced. At this time antenatal diagnosis was not possible and in any case many people in that community would have regarded abortion as sinful. About 2,300 families were seen by the researchers. In their words, the target of the programme was 'to introduce the concept of premarital exchange of genetic information into the local culture.'

Seven years later they went back to see what had happened. Virtually everyone had remembered and understood the information about themselves. However, whilst all couples where both man and woman were carriers were spot on in describing the genetic significance of their state, others tended to get it wrong. They could allow themselves to forget what it meant. But at the same time many people had come to think of carriers as having a mild disease, which is a misconception, and as intrinsically less desirable as marriage partners. Although people had tended to keep their genetic status to themselves if they were carriers, nonetheless such information was exchanged in the betrothal negotiations. Its revelation was awaited with considerable anxiety, on either side. The researchers were clearly dismayed to discover this stigmatisation of carriers, the prejudice to their chances in the marriage market and the consequent shame when marriage arrangements collapsed, as they sometimes did.

But that was not all. In the intervening seven years 101 marriages (or matings as they are called in the research report) were set up, where the individuals concerned had been through the genetic screening programme beforehand. In 4 of these both parties were carriers, which is the number that one would expect on a random basis. In 2 of the marriages one of the spouses had concealed her carrier status, and in the other 2 everything had come out into the open, and the marriage had gone ahead anyway. In other words the basic goals of the programme, to get people to exchange facts about themselves and back off, had failed. Moreover, some negotiations had failed 'unnecessarily', i.e. because only one partner was a carrier. In this case the

134

intervention turned out to be worse than useless, and its saving grace is that the principal investigator has at least said so. As I said earlier, it is not so much eugenics that lives on, but paternalism, which continues to fail, by supposing how people will act.

Times change, and with them the technology and the options open to people. By the mid-1970s antenatal diagnosis of hereditary blood disorders became possible, though not without considerable risk. In Britain cases of thalassaemia began to appear from around 1960, particularly amongst the families of Cypriot immigrants living in London. Even though repeated blood transfusions increased the life expectancy of affected children, many died in adolescence and the remainder shortly thereafter, partly because their bodies could not cope with all the iron. Eventually a drug appeared that helped considerably, and with larger doses from the late 1970s thalassaemia patients can reasonably expect to reach three score years and ten, with expert medical care and a lifetime of support from their families. This kind of help is quite expensive, way beyond the resources of less developed countries, and it demands facilities like a blood transfusion service and enough donors to keep it going. In the 1960s the birth rate in Britain of babies with thalassaemia was about 15 a year.[33] As a national percentage, that is of course tiny; within the ethnic minority concerned it would be around 1 in 170 births.

The experience of the people at risk has gone through several phases, from one of innocence where they simply went ahead and conceived without knowing the risk, to knowing the risk but knowing also that no antenatal diagnosis was possible, to a situation where it was available, and some people at least knew that this is so. Actions changed accordingly. As people came to see that they had a chance of having a child with thalassaemia, the numbers opting for abortion increased dramatically, despite community religious disapproval. In 1977 antenatal diagnosis was introduced in London and the situation changed radically. Firstly, knowledge of the disease spread through the community, partly through education and partly as relatives of affected individuals came to realise that they too might be carriers. Secondly, the rate of terminations amongst couples at risk fell from around 70 to about 30 per cent. This figure includes some spontaneous miscarriages occasioned by the procedure itself. Thirdly, the percentage of healthy children born to Cypriot couples at risk increased to 100 per cent and the rate at which pregnancies were started climbed back to roughly what it had been before people knew of the risks. Clearly then the people concerned had accepted antenatal diagnosis sufficiently to rely on it in creating a family. A very similar picture has been painted of thalassaemia prevention amongst the Italian and Greek immigrants to Quebec.[34] Amongst people from Asian communities, however, the idea of antenatal diagnosis has proved much less acceptable, and to a small number of them babies with this genetic disease continue to be born.[35] One of the factors that is crucial to the success of such exercises is consultation with the relevant community at all stages, to explain what is intended.

Our provisional conclusion must be that antenatal diagnosis works, in that it does allow people a way out of a situation that many of them find intolerable. It is also stressful and it leaves complicated feelings afterwards. It is something that people can come to accept and use, for their own purposes, if they are counselled with care and respect. But we must never assume that acceptance will be automatic because it solves what most of us think of as a major problem. Programmes of antenatal diagnosis need the most careful planning, with the full participation of the non-medical people who will be involved. They also need some kind of independent review as they continue, where people feel free to talk about all the problems and the benefits.

Chapter 9

New forms of genetic intervention

Chorionic villus sampling

The technology of antenatal diagnosis does not stand still. In the longer term it may be possible to identify foetal blood cells, which have crossed the placenta, in the mother's blood, and to sort them automatically. At the moment fluorescence-activated cell sorting (FACS) is not very reliable.[1] In the meantime a more invasive biopsy procedure that requires a small piece of tissue to be removed via the cervix is likely to have considerable impact. This is called chorionic villus sampling (CVS). It permits investigation to begin after 8 weeks of pregnancy, that is, after the woman has missed two periods. The chorion is a membrane which surrounds the foetus in the womb. It derives from the early embryo, and in one area forms villi or fingers of tissue that grow into the wall of the uterus. From these a few cells can be scraped loose for a biopsy sample. The cells obtained are genetically identical to those of the foetus, since they have a common origin.

Several years ago scientists also discovered that chorionic cells do not have to be grown in tissue culture for several weeks, but can be analysed immediately. Potentially this represents a very considerable step forward in antenatal diagnosis, because less waiting is involved and any decisions can be taken much earlier. Moreover, people often feel that it is morally more acceptable to end a pregnancy in the first rather than the second trimester, and in any case the earlier procedure is easier and safer. There is therefore considerable enthusiasm building up about CVS as a procedure.[2]

The problem is that although it seems deceptively simple to obtain

137

the sample of tissue, there is a risk to the pregnancy involved, which may be greater than with amniocentesis. Also, distinguishing chorionic from foetal cells is a special skill and adds to the cost of the procedure. CVS was first used experimentally in Scandinavia in the early 1970s, and also in China.[3] It was first used to detect inherited haemoglobin disorders in the early 1980s for women with a 25 per cent chance of having an affected child. By the end of 1984 43 centres around the world had compiled information about over 3,000 cases. In 10 per cent the pregnancy had been terminated because of the diagnosis of an affected foetus. In 4.1 per cent overall the foetus had been lost in a miscarriage after the CVS.[4] This figure includes early results as well, when the loss rate was higher. On the face of it this is about 4 times the risk with amniocentesis, but it represents intervention earlier in pregnancy when the 'background' miscarriage rate is higher, and some of these pregnancies would have miscarried anyway. This data gives an order of magnitude, but it says nothing about possible long-term risks either to the mothers or to the children. Accurately knowing the risks involved is important. One study has shown that this is the decisive issue in women's preferences for amniocentesis or CVS.[5]

This represents a serious dilemma. On the one hand it is important to assess the risks accurately and it seems unethical to carry on while only vague figures are available. On the other, setting up a meaningful comparison requires some women, who will probably be very anxious to have a child, to offer their pregnancy as a test-site. This is quite like the plans to give vitamin supplements to prevent neural tube defects. As in that case the problem is informed consent. How do you advise someone that running what could be a slightly higher risk is the right thing to do because of the knowledge that will be gained?

What is now being planned is a randomised controlled trial that seeks to compare the risks of amniocentesis with the risks of CVS, both in the short and long term. Just collecting the numbers of miscarriages, as is presently being done, is not enough, because we do not know the number of miscarriages that would have occurred in those women anyway at that stage in their pregnancy. To assess the risk due to CVS one has to make a comparison. The question is how do you create the groups of women to make a valid comparison, that is informative both about risks in pregnancy and about longer-term effects.

It seems essential to consider women who would be considering antenatal diagnosis anyway. This means that the study is likely to be based on women over 35. They form a much larger group than those at risk of any one single-gene defect, who might be pregnant at a particular time. However, they will have a higher rate of spontaneous miscarriages than would younger women, which complicates the picture. One possibility might be for women themselves to choose which procedure is used. But there is an argument against this. Women in the trial could well select themselves into two non-comparable groups, those with memories of complications in earlier pregnancies being more likely to opt for the procedure of known risk, i.e. amniocentesis. One group would then contain more women who had had obstetric problems in the past, for whatever reason.

Another possibility might be to construct a control group of women who have missed or refused amniocentesis; but again this seems likely to invalidate the comparison. For example, those who miss amniocentesis are more likely to be working-class women or from an ethnic minority, with a less frequent involvement with hospital-based antenatal care. This is not to say this is good or bad, but it does mean that there would be uncontrolled differences between the two groups of women. So what is proposed is an international collaborative study involving women over 35, in whom a viable pregnancy has been confirmed by an ultrasound scan. They would be asked to consent to random allocation into one of two groups, one having CVS at 8 weeks and the other amniocentesis at 17 weeks. The argument goes that only in this way can one compare like pregnancies and like offspring with like pregnancies and like offspring.[6]

Obviously there are major moral and psychological dilemmas here. They are not easy to resolve. For example, all the women would be asked to consider the possibility that genetic investigations would be delayed until half way through the pregnancy. But they would not themselves choose whether this occurred. It would be a matter of chance, built into the organisation of the study. Equally if the risks of CVS are significantly greater then it is possible that some women would lose normal babies after CVS, which they would probably have carried to term with amniocentesis. On the other hand, even if they are rather less, if a miscarriage occurs after CVS, there is a possibility that it would have occurred anyway, but this could only be determined by examining the foetus – in itself not an entirely pleasant thought, but something essential to the study. As things stand it seems that the study requires women to take in and digest a remarkable number of considerations and cope with their implications throughout the pregnancy. Those who participate are likely either to be people who are either ready to agree to any suggestion from their doctor or who are particularly resilient in being able to handle all these manifold uncertainties.

This kind of situation is utterly characteristic of high-technology medicine, where risk evaluation needs large carefully organised trials, but where the public and the medical profession basically mistrust each other. Medical opinion says do a trial quickly, so that we know where we stand; public sentiment is that this is using women as guinea-pigs, because they will not be told enough. The easy answer is to say that the whole thing is unjustifiable, and ought not to go ahead. But if we stop such trials, CVS might be used anyway, without any real idea of how risky it is. This is quite likely. It is what happened with amniocentesis. Or CVS will simply stop and antenatal investigations will continue to be done at 17-20 weeks, when some people will know that they could be done earlier.

Comparative trials of CVS should go ahead, but only after very serious consultation with a wide range of organisations, able to speak about the experience of genetic diagnosis and the problems of acting as a subject in a research programme. This should include self-help groups with a specific interest in genetic disease and those who deal with the trauma of miscarriage and those women's organisations campaigning for improvements in maternity services and a greater involvement by women in medical decision-making. What must not happen is that the medical profession and medical researchers take it upon themselves to decide how the study should be organised, and what people are likely to feel. Those who argue that these trials must be set up quickly before the general public has really heard about CVS undermine the whole idea of informed consent in the first place.[7] What is at stake is not just consent but also support. If people feel that their needs and anxieties throughout the pregnancy and beyond will really be noticed they will take part.

The significance of all this is hard to overstate, because of the possibility of enabling genetic intervention earlier in pregnancy. Whilst that is generally desirable, for the reasons mentioned above, the expansion of CVS could produce new problems. One is that if demand for CVS increases, as it is expected to do, the public health service will be unable to cope with it, unless genetics departments at regional centres get more money. At the moment they claim to be under-funded.[8] Another is that detection of pregnancy has to be fairly quick, which means that all concerned have to be very efficient. That in itself could be problematic, since this technology helps those who can act swiftly and leaves the rest behind.

Reading the DNA text

The antenatal diagnosis discussed in the previous chapter is based on studies of chromosomes (Down's syndrome), of a protein made by foetal cells (neural tube defect), and of foetal blood cells (inherited blood disorders). Diagnosis of other conditions is often based on levels of a particular chemical in the amniotic fluid. Within the last ten years it has become possible to obtain genetic data about the foetus by analysing very small amounts of DNA, either from blood samples or by chorionic biopsy. We can probe the genetic programme itself and inspect sections of the DNA text in which it is written. Amongst other things that means that changes *within* a particular gene can be spotted and that information used diagnostically.[9]

When you start to construct a molecular map of various genes you discover a remarkably complicated picture. In the 1960s scientists discovered that within DNA specific sequences play a whole variety of functions – stop signals, binding sites for regulatory molecules, genes that are essentially control switches that regulate how much of a particular protein is made – and so on. To add to the complexity, in the late 1970s it also became clear that genes in higher organisms are broken up into sub-sections, separated by inserts, which have no apparent textual meaning. We now perceive that there are many more places where a textual error can throw the whole system built around a particular gene into chaos. Control of the rate of synthesis is also vital. This is well worth remembering in the next chapter when we consider techniques that aim to introduce new genes into cells that lack them as a form of genetic engineering.

The problem, then, is to find some way of picking up significant variations within a text of extraordinary complexity – proof-reading at the molecular level, often without knowing what the correct version of the text should be. There are several approaches to this problem. Obviously once you know what you are looking for, it gets easier.

In the mid 1970s a class of enzymes was discovered called restriction enzymes, that had the interesting property of snapping DNA molecules at specific locations, defined by a particular sequence of bases within the DNA. If a particular enzyme was added to a test-tube of DNA molecules, the end result would be a whole lot of little pieces of DNA because particular restriction sites occur many times within each chromosome. They may, for example, be fairly close to a particular gene, so that they are almost always inherited with it. Or they may be disrupted or deleted if a section of the gene is missing. In other words, they are an indirect source of information about variation within the gene, provided they are in the vicinity. If a particular restriction site is absent, you will get bigger fragments of DNA from that particular region of the chromosome and this difference can be detected. Variation in biology is often called polymorphism and this particular kind of difference is called restriction fragment length polymorphism. But the most useful thing is that we do not have to know much about a gene to be able to use a restriction site as a marker for it. All we have to know is that they are close together, which we can infer from studies which show whether they are inherited together in a very high proportion of cases.[10]

A good example of the use of genetic markers is the recent work on Huntington's Chorea. This is a dominant condition, first characterised in 1872, which leads to a gradual breakdown in brain function, with increasingly disorderly movements and speech and dramatic changes of behaviour. This is serious enough, but the onset of the disease, and eventual death, occurs only in middle age, by which time the man or woman concerned is likely to have children and perhaps grandchildren. Until recently the only diagnostic test was the appearance of these major symptoms. We knew nothing about the gene involved, except that it is dominant. This placed the families of affected individuals in a desperate position, as they had no way of knowing whether they too would be affected. Since it is a dominant condition, the chance of inheriting the gene from an affected parent is 50 per cent; the chance that grandchildren will be affected is 25 per cent and so on.

This creates major uncertainties not just for immediate relatives, but for other branches of the family. Without a predictive test for the disease or antenatal diagnosis people had to balance the risks of inheritance against the thought of old age without children, knowing then they could have had them, or the knowledge of the distress the disease causes against the possibility of effective treatment becoming available in the interim. There are about 6,000 sufferers from the disease in the UK, many of them in particular areas in South Wales, East Anglia and Scotland, and a further 50,000 people are at risk; that is to say there is a non-trivial probability that they also carry the gene.

One research strategy here is to look for some genetic trait that turns up along with the HC gene. Over the years many have been checked and found wanting, because the association was too unreliable. Then in the early 1980s the search turned to the use of restriction sites. Several hundred have now been identified in human DNA, so that looking for a restriction fragment length polymorphism that is linked to the HC gene seems to imply a lot of false starts. However, an American group reported success in 1983.[11] They gathered together blood samples from related individuals from two very large families, one in America, and the other in Venezuela. Within both families cases of Huntington's Chorea had occurred and could be traced back through the generations for over a hundred years. In Venezuela all the individuals are part of one vast family on the shores of Lake Maracaibo, descendants of a German sailor who carried the gene.

James Gusella and his colleagues were working with a collection of fragments of DNA generated with specific restriction enzymes. Quite quickly they discovered that a rather large fragment, which they could recognise by matching it with a piece of DNA from their collection, was generated from the DNA of affected individuals and their relatives with something like the frequency that one would expect if it were inherited with the Huntington's Chorea gene. In other words this fragment seemed to be terminated by a restriction site that was a marker for the gene. Thus in principle one could then analyse the DNA of individuals at risk for Huntington's Chorea, using this polymorphism as a test of whether that individual had inherited the gene.

In practice things are not quite so clearcut, because the marker is only somewhere near the gene and the DNA fragment concerned is large. The gene has yet to be pinpointed exactly and the probability that the marker will be inherited with the gene is not 100 per cent. The closer it is the more probable it is that the two will be inherited together. This result then really focussed the research, rather than being definitive.[12] The goal is firstly to find a marker that is more reliable, so that it is really useful in antenatal diagnosis and in telling people whether they will in fact develop Huntington's Chorea. To be 99 per cent certain is valuable, but to be 85 per cent certain may not be an improvement over 50 per cent. At the moment one of the problems in Britain and usually in the United States is that families tend to be too small for it to be possible to make the kind of crossword-puzzle type inferences that are necessary. Without more generations and more sibs there are too many missing steps in the reasoning.[13] Secondly, now that the gene has been pinned down to a region on Chromosome 4, it will soon be actually identified and its DNA sequence determined. That would allow one to look for the protein specified by the gene, which would be very helpful in trying to find what causes the disease and how to prevent it.

Linkage studies like this one can represent a considerable step forward, but their intrinsic indirectness creates its own problems. For example, they require the involvement of relatives. In this case the difficulties are compounded by the specific problems of Huntington's Chorea. A predictive test for the gene, when it is available and its reliability confirmed, will allow approximately 50 per cent of those who presently know that they are at risk to breathe a deep sigh of relief, for themselves and their children. But those who are revealed as carriers of the gene will know that they will fall ill in their 40s and 50s. People's reactions to the prospect of knowing one way or the other differ considerably.[14] Some say that they want to know one way or the other, so that they can plan accordingly. Some prefer just to wait in ignorance. But confirming one's own genetic status has immediate implications for one's children, possibly with children of their own, who may not want to know. Equally, one could imagine the opposite situation where let us say a man was told that he carried the gene, which confirmed that his father was affected too, but the older man did not want to know. The morality of disclosure is not obvious here. What is vital is that those receiving this information have access to good advice and long-term psychological support, of a kind that the self-help group dealing with Huntington's Chorea, called Combat in the UK, tries to provide.

Secondly, there is the question of reproductive advice. Some people have tried to have themselves and their children sterilised. This indeed was a particular enthusiasm of the eugenists who studied Huntington's Chorea earlier this century. Requesting sterilisation for oneself is acceptable: imposing it on one's children certainly not. Until recently people have been encouraged not to have children. Most people follow this advice, although some do not.[15] In some cases the advice has been fairly coercive.[16] Some of the doctors concerned with the condition have taken an explicitly eugenist attitude towards the reproduction of those at risk, advocating very frank encounters with potential parents, to get them to remain childless.

Where inheritance is through the father, artificial insemination has been a possibility. Now we are in sight of antenatal diagnosis. One could now say to people that a wait of a few more years would be worthwhile. I believe that it makes sense not to have children in this situation. At the same time I also believe that people have a right to non-coercive advice, that recognises all aspects of their anxiety in such a position. Nancy Wexler's writings about Huntington's Chorea reveal the considerable complexity of people's responses to the risk of having inherited the gene.[17] Simply exhorting people to see reason and contracept for life, without recognising their sense of loss, their

144

capacity for denial, the psychological plasticity of the age-related risk figures and the continual return of repressed feelings, is neither helpful, humane nor effective.

Thirdly, there is the question of how research results are described by researchers and interpreted by those at risk. Claims that genetic markers have been found for some diseases have been made before. It is important to get the prospects rights. Gusella has recently been criticised for not releasing copies of his DNA probe for use in antenatal diagnosis. His counter-argument is that it is still too imprecise to be used for this purpose, and that more harm than good would be done, if people went ahead. He is probably right, since what is essential is that people have access to support and counselling, and this is not always available at present.[18]

Another recent discovery in genetics also opens up new forms of diagnosis and new dilemmas. Alec Jeffreys and his colleagues have discovered a short sequence in human DNA that is repeated several times.[19] It actually occurs as an insert within a gene. The whole unit is called a 'mini-satellite'. The number of repeats turns out to be very variable. The mini-satellite is thus a bit like a signature or a code number. It is inherited in a Mendelian fashion, i.e. as if it were a gene, so that related individuals are likely to share the same pattern of repeats within the mini-satellite. If one can find several of these and follow their re-appearance in the next generation, by comparing the DNA, you have a very good indication of whether two people are closely related.

This work has already been used in a case of disputed maternity, where a Ghanaian woman, already in the UK, wanted to bring a son back into this country. The immigration authorities felt that the child could be the son of a relative, and not the woman's. A London law centre contacted Jeffreys and he was able to confirm the woman's claim, since the mini-satellite sequences matched so closely that no other conclusion was even remotely plausible. This was done without analysing the paternal DNA. The method is now being patented.[20] The laboratory concerned has been inundated with requests for help and the method is likely to be used in other immigration cases, by those who can afford it. It will also be used in forensic science and in studying the inheritance of traits in families at risk for a particular condition.

So far we have just considered the fact that there are features within the DNA that can be recognised. We have not considered how you actually identify them. This involves the use of what are called DNA probes. The basic idea uses the structural chemistry of DNA. Chromosomes are long coiled chains of DNA. These chains are double-stranded and wrap around each other, to form the celebrated double helix. The two chains at the outside of the molecule are held together by relatively weak chemical bonds that span the gap between them. The units linked in this way are called bases and there are four of them in DNA – adenine (A), thymine (T), cystosine (C) and guanine (G). Structural and physical constraints dictate that A only binds to G and G to C. The consequence is a constant complementarity along the DNA. The sequence of thousands of bases on one strand is mirrored by the complementary sequence on the other one. Complementary pairing of bases also means that a piece of DNA with a known sequence will locate the complementary sequence, if it is around, and as a kind of molecular probe. If you label the probe DNA with a radio-isotope, you can keep track of it.

At one extreme you can use a very large probe to pick out long sections of DNA. This is what Gusella did when searching for the Huntington's Chorea gene. Or at the other extreme one can be very specific. One can look for variations in the base sequence within a particular gene. In this case one works with very short pieces of DNA made to order to identify particular re-arrangements in a gene, if its organisation is already very well known. This is an extraordinary degree of sensitivity, given that each human chromosome contains several hundred thousand bases. It is like having a test that will identify just one misprint in a book. If the change concerned is very specific, like that for sickle-cell anaemia, then this is very useful.

However, in some cases, like thalassaemia, the range of possible variations is so great, running into hundreds, that this kind of virtuosity is really no help. There are just too many different genetic mistakes that all lead to the same basic health problem. Studies so far show that most carriers within a particular national sub-population (e.g. East African Asians in the UK) have the same genetic variant. But the remaining 10 or 20 per cent may have any one of a number. Offering them antenatal diagnosis has to take account of that fact. To test for every possibility would be impossibly expensive and time-consuming. The only alternative is to use traditional methods of testing foetal blood, which again means mid-trimester investigations.[21]

Nonetheless, it would be hard to exaggerate the potential diagnostic value of DNA probes, which will mean that antenatal diagnosis is possible for a much wider range of conditions than at present. This is likely to be helpful particularly for couples who have already had an affected child. The assumption is, as ever, that people must have access to counselling, continuing psychological support and a full explanation of what the procedures imply. As I said in the previous chapter, the quality of care is defined by the number of choices open to people and the extent to which they are helped to feel they can handle those choices, not by the number of problem cases prevented. Viewed from that perspective, the mere availability of DNA probes is only a small part of the picture.

Genetic investigations in embryos

The new tools for reading the DNA text also open up the possibility of new forms of antenatal diagnosis even earlier in the life cycle. As we saw in Chapters 3, 4 and 5, embryos are more robust than one might expect. They tolerate fertilisation and early growth in a culture medium outside the body. They can cope with being frozen and thawed out. They can even compensate for the loss of one or two cells and go on to develop normally. We also know that sometimes they divide into two, or even more, so that identical sibs are born. For a hundred years embryologists have been dissociating early amphibian embryos into their constituent cells. Recently this has been done with mammals, principally sheep, to create a set of identical offspring.[22] This procedure seems to offer a whole new strategy for antenatal diagnosis, which we should take very seriously.

Embryos could be created by in vitro fertilisation, but divided up at the 2 or 4 cell stage. One part would be frozen and the other allowed to develop a little more. This one could then be analysed using DNA probes for a particular gene. Alternatively, its sex could be determined in other ways. Thus, for example, people with a risk of passing on the gene for Huntington's Chorea could find out if the embryo had acquired it. If not then the other part of the original embryo could be thawed out and transferred to the woman. As far as I can discover, this has not been attempted to date, although some work is going on to examine chromosomes in sperm, ova and early embryos.[23] The possibility has been seriously discussed by embryologists and geneticists, as something that is probably feasible and that would be desirable.[24]

From one point of view there is an obvious objection to it. Those who continue to hold that an individual human life, which is ipso facto sacrosanct, begins at fertilisation would have to condemn this, even

147

though (or perhaps because) the procedure itself weakens the very idea of individuality. For in this case one 'individual' is divided into two. One of the two parts is then used as test material to assay the status of the other. In effect a whole embryo (or part-embryo) is used as a biopsy sample. One 'individual' is created simply to see whether another shall live. However, if one denies that pre-embryos have a right to life, as I do, then there is nothing morally objectionable here. Biologically it is not significantly different from chorionic biopsy, except that implantation has not occurred. Similarly in Chapter 4 we considered the possibility of not implanting embryos but using them for research. This is basically the same.

I could imagine that someone might argue that the 'test' embryo should be transferred to someone prepared to accept the risk of having an affected child, just as I have heard the argument that the transfer of an embryo would be preferable to its destruction in the case of an ectopic pregnancy. But this argument is a non-starter. For surely very few people would in fact contemplate this – either as donors or recipients – and how would you explain it to whatever children resulted?

There are more cogent objections to the procedure. They are best considered in a moment, after some arguments for it. This procedure might seem preferable to the forms of antenatal diagnosis we have considered above. The intervention occurs prior to pregnancy and implantation. (On the other hand, the risks to the woman are the risks of ivf as a procedure, and not those of amniocentesis and abortion.)

But the probability of having an unaffected child at each attempt can only be comparable with the success rate of ivf and embryo transfer at the moment, at best, which is between 10 and 20 per cent depending on the number of embryos available. In other words the people concerned have exchanged the uncertainty in a established pregnancy for uncertainty about whether a pregnancy will be established. It is not obvious that this is an improvement. Moreover, it is only really applicable to single-gene defects, either because the problems only appear later anyway with neural tube defects or because the risks of an affected child with Down's syndrome are so low relative to the chances of success. With genetic disease the comparison is between a 25 or a 50 per cent chance of having to terminate the pregnancy on the one hand and, let us say, only a 10 per cent chance of establishing one that goes to term on the other.[25] In both cases there is the prospect of having to go through it all again, although with ivf and embryo division that prospect is several times more probable.

148

This form of pre-implantation intervention completely transforms the diagnostic procedure. The challenge of abortion is removed, at some financial cost, since ivf is not cheap. But the greater price is the uncertainty about the implantation and development of embryos being transferred after inspection. The locus of decision-making has also shifted slightly, from the woman or couple concerned, who can decide whether a pregnancy continues, to the doctor or scientist who selects the embryos. My feeling is that CVS, if it can be done safely, is preferable, because it means less extended dependence on experimental medicine and is thus more likely to enhance people's confidence that they have acted autonomously and responsibly.

Genetic screening

Antenatal diagnosis is a kind of baseline, morally, psychologically and practically. Other forms of genetic intervention nowadays presuppose it. They involve elucidating the facts of people's genetic status, in order to help them act later on that information. The action foreseen is usually antenatal diagnosis of a genetic disease. Thus only if we feel that antenatal action is tolerable should we proceed to other population-wide programmes of genetic screening designed to generate such information. Historically this has not been the order of events, but it should have been. If you give people information about themselves which they are not free to use in ways they find acceptable, you have not helped them and have struck directly at their self-esteem and self-confidence. The basic question that should be asked of most genetic screening programmes is whether they empower people to act and to anticipate that challenge without undue anxiety.

Screening can mean several different things. We have already considered some of them, although I have chosen to refer to them as antenatal diagnosis. AFP screening is one example, but here it is the particular pregnancy that is under examination, not the individual, in most cases. The same is true of amniocentesis to detect Down's Syndrome, even though this does identify some individuals with a specific predisposition to have an affected child. Also the programme of antenatal diagnosis in London for Cypriot and Asian mothers is in effect a screening programme, since it offers this help to everyone from these ethnic groups who is referred to that clinic. Another example is the blood test done on all newborn babies in most industrialised countries to detect PKU. All these programmes identify affected individuals or foetuses in the first instance; only secondarily do they reveal the genetic status of the parents.

However, one can go further than this and try to identify those who carry just one copy of a recessive gene. This was what Stamatoyanno-poulos and his co-workers did in Orchemenos with their programme of sickle-cell trait screening. In that case all the carriers could do was not to marry another carrier or not to have children. Since then several major programmes have been established that try to locate all carriers of a particular gene in a region or a state. Quite soon we are likely to see more.

One such is in Latium, the region around Rome, which has a population of 5 million people. This programme has been running for ten years and it concerns thalassaemia, which is not uncommon in some parts of Italy. Its focus is primarily on schoolchildren aged 13 to 14.[26] In all, 454,143 students have been screened to date. This must mean that most schoolchildren within the region are now being screened at this age. Of these 10,214 have been identified as carriers of one of the thalassaemia variants. A further 87,092 older people, many of them relatives of schoolchildren, have also been screened, and of these 25,121 are carriers. About 2.5 per cent of the population of this region are believed to be carriers, so about half of them have now been identified. In all 222 couples at risk – both being carriers – have now been identified. Of these 126 were identified before they had an affected child. Over the last 5 years 61 pregnancies at risk have been monitored; 15 affected foetuses were aborted, 12 of them in 1983 and 1984. This is probably 2 out of 3 of all affected foetuses that would have been born over this period. This programme is affecting the incidence of thalassaemia quite considerably.

What we do not know, as in Orchemenos, is whether some marriages never took place and what effects that had. The conventional approach to results like these is to note the decline in the number of affected children being born and to conclude that all is well. It may be. But it is also important to try to find out in some detail how people feel about the screening programme, since the decline in the incidence of thalassaemia may be built on a great deal of avoidable pain. Some attempt to do this has been done in the United States for programmes that identify carriers of the gene for Tay-Sachs disease, which is only of significant incidence amongst Ashkenazi Jews. Here the results are encouraging, although the follow-up studies have been fairly superficial.[27]

But it is clear that the involvement of community organisations (youth clubs, synagogues, charities and schools) can be very important in preparing the ground and allowing people to think about and discuss the idea of carrier identification. Interestingly though, despite the success of this initiative in America, it has not been followed in European countries.

If we focus down more closely, new details come into view. A pilot-study for hereditary blood disorders was tried out in Luton in 1983, on the fifth form of an inner-city comprehensive school, with a substantial number of children from ethnic minorities.[28] Out of 124 children in the entire year, 50 were screened, although 80 were invited to meet the doctors. Of these, 10 children were found to be carriers of a gene for one of the more common conditions and one of them seems to have had difficulty accepting this information.

What use these people, or anyone else, will make of this information in the years to come is unclear. One must hope for the best. But if one reflects on the inevitable tensions in such a school, and on the pressures on young people at this age, trying to deal with their changing sexuality, it is by no means obvious that they were helped. I can only speculate about possible problems, but we should be extremely cautious about expanding screening in schools. When the cystic fibrosis gene is identified, the suggestion will be made that carriers for what is the commonest recessive problem amongst white Europeans should be identified by screening in schools. If this happened, about 1 in 22 of Britain's school students would have some new information about themselves to digest. This is not to say that they could not do this under ideal conditions. But the question is what kind of support and explanation they would get. Screening done completely from outside, by an occasional visitor, is not the best way to proceed. The eugenic or paternalist impulse, 'to get to them before they begin having sex' is really beside the point, and potentially dangerous, if the information is simply set aside.

The whole question of carrier identification (note the jargon) needs a lot more popularisation, appraisal and discussion. The evidence from the US relating to the early screening programmes is not encouraging. Around 1970 interest in sickle-cell disease increased considerably in the US. Some black activists argued that it had been ignored by the white medical profession, or studied only as a research curiosity.[29] In 1971 President Nixon included it in his appeal to Congress for more funding for health research and suggested a special programme to combat this 'black disease'. In 1972 Congress passed the National Sickle Cell Anaemia Control Act, setting aside considerable sums of money for just one of the health problems facing black people in the US. In addition various states passed laws to screen Afro-Americans for the sickle-cell trait, and in some cases this was made mandatory for attendance at school or to obtain a marriage licence.[30] It soon became clear that some carriers identified by these rapidly spreading programmes were facing higher insurance premiums or losing the chance of promotion or being denied access to some occupations. The enthusiasm for sickle-cell disease gave way to criticism of the stigmatisation of carriers, aided by long-standing racial prejudice. Studies of screening programmes still show that significant numbers of black people are tested without their informed consent and without the possibility of counselling afterwards.[31] Their genetic status is divined from their blood and they have to live with the results.

There is a major problem here. On the one hand genetic diseases that affect ethnic minorities can simply be overlooked and the related health problems treated very cursorily at best. For example the Sickle Cell Society in the UK has argued that inadequate provision is made within the health service to help children with the disease: that births are not systematically monitored, that data on incidence and needs are not collected and that there are too few advice centres.[32] By contrast, genetic diseases that affect smaller numbers of white children have greater resources allocated to them. On the other hand it is also possible to evoke interest that is unwelcome and unhelpful, backed up with legal sanctions. Some of the US screening programmes illustrate that. What is vital, then, is full discussion with particular communities to see what they feel is needed and to explain what is possible. Only then should anyone start to plan a screening programme, on a voluntary basis.

Looking to the longer term it seems quite possible that genetic screening could be done routinely shortly after birth. After all, we do it now for PKU. In some American states it has been done for sickle-cell anaemia.[33] But as the technology evolves, and the techniques become cheaper, we may see a situation where a whole set of tests are run on blood samples from newborns, to identify those who are carriers of particular problematic genes. An argument against it is the risk of stigmatising those carriers identified as they grow up and the risk of jeopardising their relation with their parents, if the latter see the carrier state as a mild form of disease. In recent years one programme of neo-natal screening for a chromosomal abnormality supposedly associated with a predisposition to aggressive criminal behaviour, the so-called XYY syndrome, was stopped after public protest.[34] The factual basis on which the claims about the abnormality rested was very shaky and scientists at Harvard and MIT, concerned that parents were being given misleading information about their children, successfully and rightly lobbied to have the programme terminated.

This is not the only place where genetic screening is a possibility in the years to come. For a number of years people have been suggesting that it is desirable in the workplace, either to identify those who are said to be 'hypersusceptible' to the toxic or carcinogenic effects of some chemical used in production, or to pick out those who are likely to fall ill well before they are due to retire.[35] The first of these is fairly easy to evaluate, the second less so.

Hypersusceptibility screening should be resisted. Firstly, the scientific evidence on which it is based is surprisingly small and much of it worthless.[36] It survives as a hypothesis masquerading as fact. Secondly, the idea assumes that this kind of susceptibility or sensitivity to a particular chemical is very largely genetic. But other factors, such as diet, age, general health, and stress may also play a role. What I called earlier the 'implicit environment' comes into the very concept, since it brackets out any environmental complications. It simplifies both the environment and the susceptible individual. Thirdly, the assumption is that if those who are supposedly hypersusceptible can be removed from the workforce then the workplace will then be safe for those who remain. This is extremely dubious. Present trade union attitudes, particularly in the US, to criticise and oppose any attempts to do genetic screening to identify workers at risk are very sensible. The evidence suggests that employers recognise the limited value and controversiality of such tests, since the interest in them has declined in recent years.[37]

This is a short-term solution to the problems posed by particular companies contemplating screening their workers for supposed hyper-sensitivity to certain chemicals.[38] In the longer term we have to decide how to deal with the growing number of genetic markers for common conditions like cancer and heart disease. It now seems that some people are more likely to develop specific forms of heart disease, because of an inherited suceptibility.[39] They are not necessarily condemned to develop problems, if they modify their diet, but they do have a reduced capacity to deal with fat, which means they must take trouble to stay healthy. There is also evidence that some people are more likely to develop specific cancers. Regular checks may reassure

153

them if useful action can be taken. If there is nothing much that can be done, or the tests are inaccurate, then matters have been made worse.

The number of known markers for all kinds of diseases is rapidly increasing.[40] In some cases the evidence is clearcut, and the causation relatively straightforward. However, even in these cases it is not at all obvious that being forewarned is useful or desirable, if there is nothing one can do to prevent the problem developing and if one lacks any indication of when it may emerge. In other cases, amongst them schizophrenia, the terms themselves are still controversial, the evidence as to heritability very mixed, and the conditions necessary for the manifestation of the problem almost completely unknown.[41] The basis for the most appalling confusion and anxiety exists here, if these ideas are used in testing. Just as with the XYY chromosomal abnormality, we need to resist any attempt to use such precarious notions clinically at this stage.

At the same time it is very hard to see how we could keep some genetic data out of employment selection. After all, access to many occupations and eligibility for insurance cover is already based on some sort of health assessment. The question, then, is what weight to give to novel genetic data in this context. We have already seen that such information can be used incorrectly, as in the case of carriers of the sickle-cell gene, who had to pay increased insurance premiums, without any justification. It is a depressing thought that a battle to establish the implications of each marker is going to have to be fought over and over again in the decades to come.

Genetic screening is going to expand, but at what rate is impossible to say. The problems with it are easy to state. Neo-natal screening identifies children who will fall ill, allowing more rapid intervention. In the past it has sometimes been done a little too quickly, with PKU for example, before all the diagnostic subtleties were ironed out. It can also disclose carrier status, which could complicate the process of attachment to a child. It might also reveal non-paternity at this time. All these possibilities have to be thought through more carefully.

Carrier detection in schoolchildren or adults does provide helpful information, but it can also induce anxiety. As ever the bottom line is confidence, self-esteem and understanding. A great deal turns on how sexuality, contraception and reproduction are foreseen, and the extent to which possible problems in all these areas are anticipated. In one sense screening reveals only some facts, that are normally hidden below the surface of appearances. They are like others – high blood pressure, low sperm count, narrow pelvis, a heart murmur – which

154

complicate reproduction, but can be handled medically and assimilated psychologically. So, too, can complications which are genetic, if the fact that people will need some help in coming to terms with this information is not forgotten. The mere availability of antenatal diagnosis at some point in the future is not in itself enough.

Even though antenatal diagnosis may only concern a small proportion of pregnancies at the moment (3 per cent now, perhaps doubling within a decade), genetic screening would increase the numbers of people who will be forced to think about it in one way or another, as possible future parents, as expectant parents, and even as potential grandparents. To complicate the picture we stand at a moment when another approach to genetic disease is just edging into the domain of practicality. This is gene therapy, which will begin as an option for a very few people and a few conditions. How it will develop is something to ponder with some seriousness.

Chapter 10

Gene therapy

Nowadays the term genetic engineering has become too general, covering procedures in a whole variety of organisms, from bacteria and yeast, to sheep and cattle, as well as human beings. It also suffers from mechanical, Frankensteinian undertones. The label that has come to be used for the activities on which we shall concentrate here is gene therapy. The emphasis is on treatment rather than on creating a master race. This is not a linguistic feint, but a morally useful move. Admittedly there is a developing jargon, of gene delivery systems, target cells and insertion sites, with which one can get carried away.

One scientist in the US, described the pause in a colleague's research insisted on by his scientific peers, said recently:

> That's just incredible, especially if you're talking about a 3- to 5-year grant. Other investigators will be ready for primate studies very soon but Friedmann is ahead. We have the vectors. He should be in primates already.[1]

This conveys with delightful clarity the competitiveness, the crassness and the careerist urgency prevalent in this field. Nonetheless talking of gene therapy directs our critical attention to where much of the action is going on, namely very advanced and specialised medical research linked to therapeutics. It makes the discussion more concrete, and not before time. For what is staggering is the speed of events and the extent to which even limited success transforms the decision environment. Work on gene therapy feeds back to the areas discussed in the last two chapters. No technique or option can serve as a baseline for very long, because too much is changing too fast.

There are four different things to consider; only one of them is imminent, as far as human beings are concerned. These are somatic cell modification, which is the procedure closest to successful realisation; germ cell modification; the enhancement of specific genetic traits and the modification of much more complex attributes, such as physique, stature, sexual allure, longevity and intelligence, that result from the interaction of many factors, some undeniably environmental and some probably genetic.[2]

Somatic cells are all the cells of the body, except the germ cells, sperm and ova. Somatic cell modification or somatic gene therapy is the attempt to correct any malfunction by changing or supplementing the genes within them. In effect the cells are reprogrammed to function properly. This is intended to help particular individuals suffering from a particular disease. If successful, it would solve their health problem, but leave their germ cells unchanged. It is only feasible as a strategy if there is just one set of cells in the fully formed individual that is the seat of the problem, and if they are open to genetic manipulation. Germ cells modification is directed at the other category of cells, and, stretching the classificatory boundaries just a bit, the early embryo as it starts to divide. In this case all the cells of the resultant individual(s) will be affected, including future germ cells. Thus the effects are not necessarily confined to one generation, but could be passed on indefinitely.

Enhancement is somatic or germ cell modification that would not be therapeutic. Instead the intention would be to augment some hereditary trait that is socially or psychologically salient. Trying to increase a child's height in one particular way, when there is every indication that he or she would be of average height anyway, might be an example. The last possibility is engineering 'designer people', of outstanding beauty, amazing physical prowess and extraordinary intellectual powers. Whilst this is a routine achievement in certain kinds of romantic fiction without genetic engineering, it is a distant possibility in the laboratory. This is because we have virtually no idea whether any of these capacities are genetic, nor, even if that is the case, how to alter the many genes involved in the right direction. I shall consider both these forms of eugenic enhancement in a brief final section. These four possibilities form a definite sequence. Being able to do one of them makes the next one easier. But the question we must ask is whether the movement through the sequence is inevitable. Do they represent a slippery slope, down which humanity will slide into moral chaos, if we allow the first step to be taken?

Targeting parts of the soma

Inevitably if we cannot modify sufficiently the interaction of the organism with its environment we are led to consider altering the genetic programme, at least in those cells where it needs to be right. Reductionist this may be, delivering a genetic 'magic bullet' to a target in the soma, but if it works it may mean the difference between life and death. Moreover, this approach has 'worked' so often in medicine and research, that it has become a kind of common sense to scientists today.

Manipulating genes is a common research strategy in the life sciences in the twentieth century. Much of the foundation of contemporary biology has been built on work with bacteria, which are very simple organisms, and viruses, which are even more basic, lacking a cellular structure altogether. Moreover, viruses can be used to pick up genes and move around with them. Many have the ability to incorporate themselves into the chromosome(s) of the cell they have entered, and be copied as part of a chromosome as the cell divides. In this passive state they are known as proviruses. This goes on until something switches the virus back into its more active form. They may also pick up DNA from chromosomes into which they have inserted themselves and act as carriers of this non-viral DNA.

In the mid-1970s molecular biologists discovered new ways of making gene transfers in a much more precise and controllable fashion. This dramatic step forward depended on an ability to isolate specific genes, using the restriction enzymes mentioned in Chapter 9, and to 'splice' them into mobile genetic elements, which were either viruses or a kind of bacterial mini-chromosome called a plasmid. More recently other ways of making similar transfers of genes have been devised that work in plants, insects, reptiles, mammals, and human beings.[3]

Three implications for dealing with genetic disease are particularly important. The first is a vast increase in our ability to sequence DNA and to examine the detailed organisation of particular gene systems at the molecular level. The second is the ability to go beyond mere analysis to make DNA sequences to order and to copy them in large numbers if necessary. The DNA probes mentioned in the previous chapter are an example of this. Now we can also think of genes as components that can be made on demand. The third implication is we can consider supplying a package of genes to cells in which a particular gene-controlled process is malfunctioning in order to put things right. Of course, this vastly oversimplifies the practical difficulties because of the complexity of the cellular and genetic systems involved, but it captures the philosophy of the approach.

This prospect came a step closer in 1969, with the discovery of classical genetic procedures that allowed the isolation of specific genes, in bacteria. On this occasion the scientists concerned, led by the Harvard biologist, Jonathan Beckwith, explicitly drew the attention of the public both to the power of these techniques and their wider applicability and also to the fact that in countries like the US many potentially beneficial scientific ideas are turned to inhuman ends by those who control science.[4] I can trace my own motivation to write a book like this one back to hearing Beckwith speak in London in 1970 on this theme.

By 1972 the idea of using sections of DNA constructed to order to correct human gene defects was being seriously discussed in major scientific journals.[5] One ill-conceived attempt along slightly different lines was made in 1970 with two mentally retarded girls in Germany.[6] By the late 1970s, with new knowledge and techniques from recombinant DNA research putting specific genes into animal cells and getting them to work, there was a feasible proposition.

Then in the autumn of 1980 came the news in the *Los Angeles Times* that an attempt at gene therapy in two human patients had been made that summer. This led to a flurry of articles on the subject, one with the revealing title, 'Gene therapy in human beings: when is it ethical to begin?'[7] As more details emerged of the first attempt, an international controversy broke as to its propriety.[8]

The scientist principally concerned in this case was Dr Martin Cline, a haematologist at UCLA Medical School. He was a clinician, with experience of the hereditary blood disorders like sickle-cell anaemia. During the late 1970s he collaborated with a molecular biologist at the same university, Winston Salser, to find ways of transferring β-globin genes into the bone marrow of mice, in the hope of being able in the longer term to do somatic cell gene therapy in human beings. In the spring of 1980 Cline and Salser published what they believed to be promising results with mice.[9]

Cline and Salser believed that their methodology could be applied in human beings suffering from thalassaemia. Blood cells are made

159

within the bone marrow. There the cells divide very rapidly and they constitute the one human tissue, apart from skin, that can be removed from the body, grown in culture, and then replaced. The idea was to try to put globin genes into human bone marrow cells and then return these cells to patients with thalassaemia, in the hope that enough would survive to raise the amount of normal haemoglobin being produced. But there were many unknowns and this would be a highly experimental procedure. Hence the work with mice.

Since the Second World War we have had second thoughts about trusting doctors and scientists to take their own decisions about what is an acceptable experiment in a human patient. But despite the horrors perpetrated by German and Japanese researchers in the war, the process of creating committees to check research protocols and to monitor standards has been remarkably slow.[10] Nowadays, particularly in the US, permission to do clinical research with human subjects has to be sought from local panels of researchers and laypeople, called Institutional Review Boards. In Britain the position is similar, although the exercise is often taken much less seriously.[11]

Cline and Salser applied for permission in Los Angeles to take their work a stage further, by using human beings. The two local committees there referred the matter to each other, neither being able to decide which body had the primary responsibility for the assessment of such a significant piece of work. All this took time. In an effort to get an answer the experiments were referred to a national committee in Washington, the Recombinant DNA Advisory Committee of the National Institutes of Health, for approval. They too referred it back to the human subjects protection committee at Cline's hospital. All this took more time.

Meanwhile, and partly because of the delay, approval was also being sought to do this work outside the US. Two different hospitals in Jerusalem and Naples eventually gave permission. In the US the eventual result went the other way. In the summer of 1980 Cline went to Jerusalem to try his experimental procedure in a young woman of 21 suffering from β-thalassaemia. She had herself consented to act as a research subject.

The understanding was that the genes would be introduced as separate pieces of DNA, and not spliced together as a recombinant molecule on a plasmid carrier. On this esoteric technical point Cline changed his mind, apparently after he arrived, because he believed it would be helpful and would have no untoward effects. However, he neglected to tell anyone of the change of procedure, which formally

160

was non-trivial. It was this undiscussed departure from the agreed protocol that got him into trouble.

When he returned to the States and news of his attempt emerged, he faced some scientific criticism that the move to human patients was scientifically premature, and was thus morally unacceptable. This led on to more national publicity, as the fact of the misrepresentation of the work came out. Eventually he resigned his clinical post and was deprived of one of his substantial research grants from NIH for this work. He retained his university post and other research funds for other projects.[12] His action provoked a vast amount of media coverage around the world, not all of it unfavourable. He has won for himself a place in the footnotes of history, though perhaps not in the way he intended. He was punished *pour encourager les autres* to stick to the letter of their agreement with ethics committees, and not because any physical harm came directly to the people involved. I think that the eventual volume of public criticism has led some scientists in the field to take his side, or to say, there but for the grace of God go I. When I heard him speak at one scientific meeting, I was surprised by the interest in the potential of his technique and the lack of criticism. The same has been true on other occasions, although other scientists, also in the same field, have clearly and forcefully said that it was wrong of him to proceed with so little chance of success, that his own experiments suggested he would fail, and that doing work abroad that already had been criticised in the US was not acceptable.[13]

Since that time there has been a great deal of discussion, mostly in the US, of the technical and procedural steps that must be taken before another attempt goes ahead. In America at least the fact that some scientists are keen to pull off this particular feat has been recognised. There considerable attempts have been made to promote public discussion of the acceptability of gene therapy and its implications. These include a Presidential Commission, Congressional Hearings, a background paper on gene thereapy prepared for the Congress by its Office of Technology Assessment and several documents from a Gene Therapy Working Group of the National Institutes of Health.[14] This makes a very striking comparison with the occasional bleats of concern from governments in other countries, followed by marked reluctance to creat any fora, public or otherwise, where such issues could be discussed.[15]

Much more scientific work has been done with various test systems using animals and human cells growing in culture. With the benefit of hindsight, Cline's actions in 1980 do look very premature. There has been a shift towards the use of specially modified viruses, as a more efficient and controllable way of getting genes in cultured cells, and away from thalassaemia and sickle-cell anaemia, where the control of gene expression is much more complex than was originally thought. Making the haemoglobin molecule is a very complex business, involving the coordinated activity of whole sets of genes on two different chromosomes, to make just the right amount of each part of the molecule. So it makes sense to work with conditions where the rate of synthesis is not so crucial. Interest has centred recently on a very rare neurological disorder called Lesch-Nyhan syndrome, which affects 200 children a year in the US (10-20 times less common than sickle-cell anaemia). Even this may be difficult to handle by gene therapy in its present form, because although bone marrow cells have been genetically engineered to produce the missing enzyme, no one knows how to get brain cells to do this, since they do not divide, and they are the cells that need it.[16]

Consequently scientists are turning to even rarer inherited immune deficiency diseases. One of these has been diagnosed in only a handful of children in the world. In such cases the body is unable to defend itself against infections and the children only survive either with extremely complex drug treatments or in total physical isolation, living in sterile plastic bubbles. These are children in a desperate plight, on whom a great deal of money will be spent, because they represent a very interesting challenge technically. The knowledge gained is likely to be of general relevance, but the same will be true of many other less spectacular problems.

A number of reports in the scientific press make it clear that another attempt at human somatic cell gene thereapy is likely in 1986.[17] The idea of placing genes into human cells to treat a disease has obviously seized biologists' imagination. Procedures involving gene transfers are central to cell biology and molecular genetics. It has become second nature for scientists to think in such terms. Somatic cell modification has become a research goal for 8 to 10 groups in the US, in a way that few would have predicted 10 years ago. Inevitably that means some competition to be the first to succeed. Gene thereapy has arrived as a legitimate and appealing idea, within the sub-culture of research, if not as a practical proposition. The question now is how soon may the next step be taken.

The pros and cons of somatic cell therapy

Is this remarkable shift of attitudes to be welcomed or attacked, given that biologists often used to say that all this lay far in the future and speculation about the possibility was pointless, if not downright irresponsible? There are two questions to consider. Is gene therapy acceptable and if so is enough known for it to be attempted in 1986?

Why attempt somatic cell gene therapy at all? Would it not be much better simply to forswear this line of research, for ever? Three considerations oppose this view, none of them overwhelming, but all important. The first is that many genetic diseases are very difficult to treat and impossible to cure at the moment. We cannot bet on therapeutic breakthroughs in other areas, such as new drugs. On this argument gene therapy increases the number of ways of trying to help people, often young people, who are chronically sick, or terminally ill. The fact that the intended therapy involves the genes is not especially important. Gene therapy is in effect just another kind of replacement therapy, like blood transfusion, or enzyme replacement, using genes rather than their products.

The second argument for gene therapy is that some people find one of the alternatives, namely selective abortion, unacceptable and that they and their children deserve some help. If one believes, as I do, that people deserve the widest range of options, when they face the challenge of genetic disease, and should not be confronted by the stark choice of abortion or nothing but criticism, then gene therapy is something worth pursuing. The same is true of those who would not have refused an abortion, but whose problems are discovered too late for the pregnancy to be legally terminated.

This leads to a third argument for the procedure. At the moment the majority of problems occur in people whose parents had no idea that this was a possibility. They did not know that they were carriers. Even though genetic screening will become more widespread for some common conditions, it is likely to be the case for a very long while, if not for ever, that some problems will remain unforeseen or undetected until birth.

Arguments like these are hard to gainsay. Only if we felt that something very precious was almost certain to be lost, soon after we accepted the practice of gene therapy, should we decline to take the next step. This in fact is just what the exponents of the 'slippery slope' argument maintain. In their view once we have begun we will career helplessly from actions with a medical rationale towards more and more trivial and self-indulgent tinkering with the human genome, into

163

the twilight world of eugenic narcissism. In the process we will have traded our humanity for a bogus ideal of perfection. The American critic of many forms of genetic engineering, Jeremy Rifkin, has argued on these lines.[18]

This argument is wrongheaded, very conservative, even reactionary, because it says so little about what we ought to do. Instead it deplores a trend, based on questionable extrapolations, and evokes a natural and Utopian state from which we depart at our moral peril. It is a philosophy of stasis and timidity. It is as if when we move from our present state of relative powerlessness, we shall lose any moral sensitivity. It is not at all obvious why this should be so.

Ironically I think these trends towards insensitivity and a contraction of our moral horizons are more likely if we do not develop various forms of gene therapy to deal with genetic disease. For if gene therapy is not developed, but antenatal diagnosis continues to be refined, then those people who have conscientious objections to abortion may well find themselves pressured to accept a procedure that continues to become easier technically. I say this despite my basic endorsement of antenatal diagnosis and the feeling that it is preferable to gene therapy in most cases. Antenatal diagnosis confronts the problem of genetic abnormality head-on, and increasingly allows people to decide in the early weeks of pregnancy what they want to do.

Gene therapy, on the other hand, probably means years of medical treatment, uncertainty about their effectiveness and anxiety about unknown risks. It is ludicrous to suggest that it will be so straightforward that people will opt for it just because they want beautiful and talented children. It is likely to remain a very tricky procedure and certainly not something one could choose lightly. Pressing ahead with the research that will make gene therapy possible will not lead us inexorably towards unthinking genetic engineering.

So the basic case against somatic cell gene therapy fails. That does not mean that we should begin immediately, or even accord it a high priority. It is simply saying that we have no need to call a halt to research in progress out of a sense that our basic values are under threat.

However, there remains the question of risks to patients. We must know, from animal studies, that the gene concerned can be put into the target cells and will stay there long enough to be effective. We must be certain that the gene product will be expressed at an appropriate level. Too little is no help; too much could be worse than useless. Thirdly, we need to be sure that the effects of inserting the gene, on the

164

chromosomes, cells and tissues involved, will not themselves be harmful.

Almost certainly specially constructed RNA viruses will be used to transfer the extraneous genes into the target cells. There are other ways of attempting this, but none are as effective or as controllable.[19] They have been used to put specific genes into mouse bone-marrow cells. This has just been done with human bone-marrow cells by Hock and Miller in America.[20] Unfortunately their experiments also demonstrated one of the major problems that still remains. The whole point of going through a very complicated rigmarole is to leave the infectious virus behind. In this case it turned up in the culture medium, which is most undesirable.

There are other problems here, too. All this is being done in culture dishes on the lab bench. Within the body the target cells are the stem cells in the bone marrow, from which many kinds of bone-marrow and blood cells derive. There is evidence which suggests that these cells have very effective ways of shutting down the expression of whole sets of viral genes, when viruses get into them, presumably as a very powerful defence mechanism. Nobody knows how to get around this problem, if it exists.[21]

Furthermore the viral vectors in use here are not yet sufficiently specific. We cannot be certain that they will only infect particular target cells. They might, for example, wander off in the bloodstream and infect germ cells. Moreover, to get them inserted cell division is necessary. How one infects non-dividing brain cells – a frightening prospect in its own right, but probably essential with Lesch-Nyhan syndrome – is not clear. However, if viruses could be made tissue- or cell-specific, if they recognised receptors found only in certain kinds of cells, then they would become much more precise and reliable tools.

Finally, these viruses insert themselves randomly into the chromosome. This is a major drawback. It is rather unlikely if they can slot themselves in wherever they see fit that the synthesis of their gene products will be correctly regulated. And worst of all they seem to cause a fair amount of disruption as they insert themselves. This is very bad news, because not only do they re-arrange themselves in the process, they also wreak havoc with the chromosomal genes at the insertion site. It is like findRNA virusing right in the middle of a word, which scrambles the meaning of the sentence. One serious consequence is cancer, which can occur when specific genes called oncogenes are switched to their active state.

The three criteria mentioned above, of delivery to specific tissues, stable expression of the gene and safety, have not yet been met, the last criterion most obviously so. It is really very surprising that with problems as large as these still unresolved there should be such scientific optimism that somatic cell therapy will soon begin. Nonetheless this message comes through in many recent review articles.

The obvious analogy here is with experimental treatment for cancer. Currently we find it generally acceptable that clinical experiments should be done, with informed consent from the patients concerned and after scrutiny by an ethics committee, on people who are terminally ill, often in great pain, and who know that the work will not save them but may provide some useful information. Apart from consent, the conventional wisdom is that the experiments should be well thought out, can be done in no other way, and will not actually make matters worse. Formally diseases like Lesch-Nyhan syndrome and severe combined immune deficiency are very similar, except that parents will have to consent for their children, as they do in leukemia. But as yet we are not at the stage where clinical experiments can begin. The worrying thing is that the competitive pressure on scientists and doctors makes some of them eager to get on to that stage. This is the most effective way to destroy the mutual trust and respect between doctors and their patient-subjects, and we need very strict review procedures, to which the public have access, to make sure that the work begins only when everyone is agreed that it makes sense.

In the US some care has been taken to create the relevant committees, and to make it clear to researchers what kind of scrutiny they will face. This may yet turn out to be Martin Cline's greatest contribution, the strengthening of the regulatory system, to cope with people even more eager to start than he was. On the current US documents prepared for the guidance of researchers seeking permission to go ahead are two questions that do not normally appear on such forms.

A. What steps will be taken to ensure that accurate information is made available to the public with respect to such public concerns as may arise from the proposed study?
B. Do you or your funding sources intend to protect under patent or trade secrecy laws either the products or the procedures developed in the proposed study?[22]

Both these are very sensible; indeed they should be asked all the time, not just in special cases. The first has two purposes: firstly, to make sure that the patients concerned will be adequately protected from

166

unhelpful media attention, in ways that are agreed with all concerned, but secondly, to protect the research from criticism born of misunderstanding. Another lesson to be drawn from the ivf saga is that not only should research teams anticipate major problems from intense media interest, but also it is very much in the public interest to ensure that the way this will be handled is agreed beforehand. There were major difficulties in Oldham in 1978 because different parties felt they had the sole right to control access to the people in the limelight. Taking such precautions in future will kill sensational reporting of medical firsts, and stop both distress and windfall gains to the people concerned.

The second question is also very serious, because the implication is that commercial considerations could stop the full disclosure of what was intended. As we have seen with ivf, and also embryo flushing, this is already a consideration in that field.[23] As we shall see in the next section, the importance of this technology in animal agriculture may mean that techniques are patented. It is vital to ensure that all the details of proposed procedures are fully discussed and all details made public. If that means that some things cannot be patented that is just too bad.

Germ cell gene therapy

Everything said so far has been restricted to work with somatic cells, usually bone-marrow stem cells, but possibly liver and even brain cells in the future. The presumption is that the direct effects of somatic cell modification are not passed on to future generations, because the extraneous genes are not allowed to get into the germ line. The individual who survives to reproduce is assumed to have unaltered germ cells. This generational barrier is usually regarded as very important, as a cordon sanitaire between our experiments and the future. Obviously there will be indirect effects on future generations if somatic cell gene therapy is begun. These are worth analysing.

One eugenist view might be that somatic cell gene therapy suspends the action of natural selection against genetically inferior individuals, who pass on their genes, when otherwise they would have died, to the detriment of the human race. This could be another argument against gene therapy *per se*, or it could be turned into a case for attempting to correct germ cells as well. All or nothing then, for the good of the stock.

Obviously it is true that if we treat individuals with a genetic disease, then they must pass on one set of their genes to offspring, if they have them. This applies to any form of therapy applied to genetic disease. The numerical consequences of this depend in the first instance on the mode of inheritance of the relevant gene and the genetic status of the reproductive partner.

But the practical and genetic significance of all this really depends on what happens when the people concerned attempt to conceive and a pregnancy begins. People who have been treated for a genetic disease will know what their problem is, and, unless they are totally alienated from the health service, they will know what kinds of help are available. We know that the majority of people with some experience of genetic disease in their family will opt for antenatal diagnosis and selective abortion. We need not then fear that somatic cell gene therapy is likely to lead to generations of people needing similar treatment. I am sure that precisely because they have survived people would not wish to put their own children through that experience. The virtue of antenatal diagnosis is that it allows them to have healthy children.

A more persuasive argument for germ cell modification in human beings has been raised already. There are diseases, cystic fibrosis being one, that affect organs or tissues throughout the body. Since the cells involved differentiate early in development, one might contemplate acting on the early embryo, on the fertilised ovum, or the germ cells prior to fertilisation.

Now this might seen a fine example of a slippery slope. I began with somatic cell modification, made something of a case for it, as well as some criticisms, and here we are contemplating the next step. Obviously there is a slope here. I am deliberately drawing attention to it. However, I do not believe that it is slippery. That is to say, we can both identify relevant moral distinctions between various forms of genetic intervention and use them very effectively to block any move to do gene therapy in germ cells. Moreover, I believe this is so despite considerable commercial pressure to develop the technology in other areas such as animal breeding.

Germ cell modification has been attempted many times with various degrees of success in laboratory and agricultural animals.[24] It has not yet been done in human beings, at least in ways that are intended to result in an implanted embryo or an infant. There are three basic reasons for such experiments. One is to find out what particular genes do, by inserting them into the pronuclei of ova, where they are

168

expressed. The second is to develop animals with a particular trait, by analogy with breeding for such characteristics. The third is to get the body of the animal to produce a particular substance of commercial value and secrete it into its bloodstream or milk. An analogy here is with the production of antisera in horses or rabbits.

The point to bear in mind is that all these applications are being pursued with considerable funds. It is not uncommon to find government or business reports assessing the prospects for the 'embryo industry', a slightly troubling phrase in itself, although it applies to animals.[25] The reference here is to the global trade in frozen embryos, particularly in dairy farming. Being able to offer for sale sexed embryos with specific characteristics is commercially very appealing. However, this must mean in the medium term that techniques for gene transfer into embryos will improve and their application to human embryos or germ cells will be discussed. Friction down the moral slope due to technical difficulties is likely to decrease.

Germ cell modification could be done in one of several ways. One is some variant of the viral methods already considered above. Of the many problems with them, a major one is 'insertional mutagenesis', the genetic disruption that insertion into the chromosome causes. Another possible method is the micro-injection of bits of DNA directly into the nucleus of the fertilised egg. This procedure is often used with animals. Probably its most striking use to date has been to make mice that overproduced growth hormone, because copies of the growth hormone gene had been inserted into the ova from which they developed. This gene was spliced to a kind of molecular switch, that could be turned on by substances in the diet. When these mice grew up they were much bigger than their undoctored litter-mates and many cells in their bodies were producing growth hormones. The micro-injection technique used here is not applicable to human eggs, because it is traumatic, because human eggs are even smaller and because many copies of the gene are introduced into the egg nucleus.[26] This technique is being used, however, to modify the characteristics of farm animals.[27] The third possibility is some sort of fusion of embryo cells or embryos. Again this has been done in animals, for example to produce interspecific hybrids, such as the goat-sheep.[28]

There are basically two arguments against the use of such procedures with human embryos or with germ cells. The first one relates to the recurrence of unintended genetic problems in subsequent generations. If the effects of gene therapy on the next generation are really disastrous, then that generation will die in utero. That will be tragic in itself. But as we know from the studies of populations affected by radiation from nuclear weapons or highly radioactive material, the genetic effects may be less drastic and take longer to be eliminated. We should try very hard indeed to avoid the serious congenital abnormalities of unknown cause that have appeared in such populations. It is perfectly possible that such problems could arise in the descendants of people subject to germ cell gene therapy, because of some unknown complication of the genetic intervention.

The other argument is that taking such risks is unnecessary anyway, if one finds antenatal diagnosis acceptable.[29] The chances of success are already high, and the medical complications not very great. But I said earlier that we should not forget that for some people it is not acceptable. However, it is very unlikely that such people will feel able to use procedures that could only be developed through experimentation on many hundreds or thousands of human embryos and which when applied must place the continued development of any embryo in jeopardy. It is not like foetal surgery where some people argue that risky procedures should be attempted to help a foetus to survive, placing the mother at some risk in so doing.

Neither of these arguments is absolute. They depend for their force upon technical unknowns. I could just about imagine that some day we would know enough to be certain that the problems envisaged here with germ cell modification would not occur. Until that time the consensus must remain secure that such experiments are completely unacceptable. This is the case at present. Legislation or very clear policy directives to scientists to reinforce that position would be very helpful.

Designer people

Finally, there is the prospect of tinkering genetically with various bodily or mental characteristics, to create offspring who are just that little bit smarter and more in control, who are *really* charming, who look so good and who will outlive us all. If this sounds familiar, it should do; these are the people from the adverts, the designer people with style built into them.

The only effective attack on this silliness is through satire, rather than dispassionate reason. I shall leave that to someone else, if Aldous Huxley has not already done the job adequately in *Brave New World*. In fact he did not say it all, for he concentrates his attack on the

170

blandness and inflexibility of a consumer society created and maintained by centralised state activity. The very idea of parenthood and family life has been abolished. Today it is rather more apparent that we should be on guard against entrepreneurship that seeks to capitalise upon desires within potential parents.

The excesses of the in vitro fertilisation business, the surrogacy industry and the market in sex predetermination services all arise because parents' anxieties are perceived as profit opportunities by people with a skill to sell. The commercialisation of human genetics could easily take the same form. In fact it is slightly surprising that the development of private genetic services in the UK has been so slow. In other countries like the US where health care remains a matter of private consumption, except for people whose purchasing of medical services is done for them, it is already the case that what you get in the way of genetic assistance depends on how much you can pay. The question of which antenatal investigations should be included in standard policies is now being debated by the insurance industry in the US.

The enhancement of genetic traits and full-blown genetic engineering are also about realising parental hopes for their offspring. If and when such things become possible, people might be persuaded to pay for them. But here we have moved on from therapy or diagnostic investigation to deliberate design. The question to consider before these things are possible is whether such desires are legitimate, and if not, whether we can act against those who try to service them, as we have already done in the UK with commercial surrogacy.

Technically there are considerable differences between enhancing characteristics controlled by single genes and engineering more complex traits. Morally I would classify them together. Targeting a package of genes into specific tissues to enhance a specific function seems a plausible possibility. It is not possible at the moment with human beings, although we have already seen that it can be made to work in mice. In that case, however, the added genes are active in many places where they should not be – in the testes and muscle, for example. In time no doubt this technology will improve, and its use in human beings for non-therapeutic purposes might well be contemplated. An example that has been discussed in the biological literature is increasing the production of growth hormone within the body.

Engineering much more complex traits or qualities, such as athletic performance, longevity, intelligence – whatever that is – body shape or temperament is a different ball game. This is both because these things are only poorly defined anyway and because the little that we know about them so far indicates that their variation is affected by some *interaction* of environmental and hereditary factors. This means that we have no idea at present what such intervention would achieve. Despite these technical differences, in one sense both these modes of intervention are similar morally. Both represent ways by which parental hopes for their children might be realised, assuming that the state and obligations of parenthood are not abolished as they had been in *Brave New World*. But both place their well-being and development in some jeopardy, which is why I consider them together. The risks involved are twofold. There may be medical problems, since we have very limited understanding of the systematic effects of trying to enhance the functioning of a gene that is already working normally. And there may be psychological problems, if children fail to live up to the expectations of the design to which they were built.

Thinking in this way allows us to construct analogies with existing situations and to form a response. We allow parents to buy special education for their children, although many people, myself included, object to a stratified market in educational opportunities. A closer analogy would be with the purchase of black-market growth hormone to be injected into children of normal stature, in the hope of giving them an advantage as basketball stars or athletes. Now that growth hormone can be made cheaply in bacteria this practice is starting in the US. It is clearly wrong, because of the many medical unknowns. Genetic enhancement of growth hormone production would be very similar, given the present state of ignorance. Indeed it would be worse because the effects could not be stopped.

One can also condemn the engineering of designer people on similar grounds. It may be of course that I am simply reflecting the prejudices of an academic in a market culture. Not everyone, perhaps, will share my distaste for the introduction of brand-names or product images into the human germ-line. We still talk of 'blue blood', meaning blue genes, or 'good stock' and so on, and even though these terms have no scientific meaning, some people apparently feel they denote something tangible. Soon we shall be able to speak more precisely about gene-clusters that seem to correlate with desirable physical and mental attributes. Why not market them in the idiom of the age, as designer genes? What would the objection be to what has been called the 'genetic supermarket', where one could select packaged attributes or individualised embryos according to one's preferences and ability to pay?[30]

Perhaps there is no ultimate or overwhelming objection of the kind suggested by the now rather over-worked term 'playing God', which is always used to imply that we have gone much too far? But two less grandiose objections are for me quite decisive. Firstly I reject the reductionism implicit in the idea that human qualities like intelligence or beauty can be thought of as traits that are either present or absent and can be produced by the simple reprogramming of the embryo. This is a desperate oversimplification of human complexity and human variability. If it were not so neatly compatible with the ideology of consumer society – you are what you have bought – you could have a good laugh at it. This leads to my second objection, which is to the penetration of a certain kind of consumer desperation and anxiety into procreation itself. Not content with buying special education, not satisfied with the mutual support of other expectant couples in pregnancy classes, not reassured by the idea of pre-conceptual care, some people apparently would want to go even further and buy yet more options on the future, by picking their embryo in the supermarket, designed for them by the great people-designers of our time.

I do not believe in a human essence, made in the image of God. I have no theological objection to the idea of modifying conventional human characteristics. However, I do think we can lose the possibility for real self-understanding if we cave in to such crass commercialism, if we are persuaded to think of ourselves as sets of components that we assemble according to the latest style and within the constraints of our household budgets. Designer people after all are condemned to live as mere advertisements for their genes, blue or otherwise.

Chapter 11

Back to nature

Defining needs

In the 1980s we see innovation as the business of entrepreneurs. A great deal of this book has been about entrepreneurial activity in science and medicine. It is a treatise on how best to respond to research and commercial action designed to create, manage and profit from new opportunities for the consumers of medical services. Virtually everything discussed here can be seen from this perspective.

In the eighteenth century John Hunter built up a substantial medical business in fashionable London society. With great discretion he advised one of his clients that insemination could be effected with a syringe. In the twentieth century sperm banking grew to meet the desire of US conscripts that some of their germ cells should survive a tour of duty in Indochina. Ivf has burgeoned as entrepreneurial obstetricians have come to see it as a valuable addition to the package of services they can offer their clients. Sex predetermination in India, and elsewhere no doubt, is made possible by doctors or scientists selling a technical skill to fathers desperate not to have a daughter. Surrogacy is par excellence a service marketed to people whose needs for children apparently cannot be met in any other way. Genetic intervention too is sometimes reassurance for a fee.

This way of describing things is easily misunderstood. It may seem that by referring to money, consumer demand, marketing, entrepreneurship and profit I am both insulting the medical profession, by implying that they are all motivated simply by financial greed, and patronising their clients as helpless dupes. I do not have such a simple-minded and misanthropic view of either party.

Yet this interpretation, applied selectively, cannot be rejected entirely. Some doctors' and medical researchers' interest in money is undeniable. They make their research and career choices accordingly. It would be naive to pretend otherwise. In this book I have discussed a few cases where this has obviously occurred, including those where new technology (embryo lavage) or a new concept (commercial surrogacy) has been backed by venture capital seeking a profit on an initial investment. I also think that some people have bought services that really they did not need, with psychological and medical consequences they will later regret. People poorly advised by private ivf clinics are one example. Moreover, in the chapters on surrogacy, sex predetermination and genetic engineering I have argued that people should be stopped from acting on some of their desires, although some will seek to help them, principally because they are likely to do harm to others in the process.

But my primary intention in framing things this way is to draw attention to the fact that anxieties and medical problems are constantly being turned into 'consumer needs' and entrepreneurial opportunities. That is what biomedical innovation is all about, for good or ill. If you ask doctors and scientists why they are promoting ivf or gene therapy, the answer comes back: because people feel they need it. This is undeniably true, but not the whole story. This answer only re-directs us to the question of what actually happens for people to feel something as a need.

All the things discussed in this book relate to urgent problems, or in the case of sex predetermination to matters that excite strong feelings. By using the langugage of entrepreneurship and consumption, I am not implying that preferences here are just like those for spin driers or compact disc players, which are fairly trivial in the context of someone's life as a whole. The point is rather to stop ourselves from saying that certain desires and wants are only 'natural' and from leaving matters there. Even though the desire for children may possibly be innate, or the feeling that one could not cope with a handicapped child deep-seated, the ways in which we develop these feelings and state what we think of as our needs depend upon culture. How we translate very basic hopes and fears into the language of needs is a product of social experience. So too is the collective judgment of whether and how such needs should be met.

Thus ivf is expanding firstly because medical entrepreneurs perceive that a demand for it exists, and are prepared to invest time, energy and money in its development, secondly because some infertile couples feel it speaks to their need for children, and thirdly because society at large sanctions both this expression of a need and the means selected to satisfy it.

In fact of course matters are nothing like so clearcut or decided. I have been writing this book precisely because controversy continues over many reproductive technologies and whether the needs to which they are addressed should be given attention. For example, some people would say of ivf that infertile couples ought not to feel such a strong need for children and should be helped in other ways to come to terms with infertility. Others would emphasise that doctors ought to put their efforts into satisfying other needs, such as better preventive medicine. Others again, whilst acknowledging that some people are frustrated in their desire to have children and that ivf could help, nonetheless feel that society cannot condone the practice and the research with which it is associated.

It is utterly characteristic of disagreements in this area that critics often find themselves saying that people ought not to feel a particular need. To press this line of argument seems extremely harsh. To withdraw without further discussion seems vacillatory and empty-headed. The only way forward is to consider how the situation arose. This has to be a continuing undertaking that will run alongside biomedical research indefinitely, as a constant commentary.

I have tried to provide some thoughts on how particular areas of medical science and technology have evolved, but a great deal more work needs to be done. For example, we know very little about how medical researchers develop their careers, and what makes particular projects or perspectives appealing. The best example of this is the upsurge of interest in gene therapy, which has clearly seized some scientists' imagination at the present time. Similarly, the way in which particular areas of work suddenly become controversial is largely mysterious. Most evidently this has occurred with research on human embryos. Here the mere fact that certain experiments are now technically feasible is nothing like the whole story. Also infertility has clearly been 'discovered' as a problem within the last few years. This is partly related to the simple increase in incidence, but again that is only a small part of the process by which certain forms of medical help like ivf have come to be seen as meeting a need.

176

Dissent from the path of high technology

One common response to the problems discussed in this book is to say that the technology should just be set aside, or left undeveloped. This attitude has been expressed both on the Left and on the Right. In each case commentators have evoked images of natural simplicity or a sense of union with nature as an alternative to the application of reproductive technologies. Nature, then, is a symbol of technological dissent, and a source of imagery and rhetoric with which to argue that needs have been wrongly formulated or will be met in dehumanising ways.

In April 1985 a congress was held in Bonn of 'Women against gene technology and reproductive technologies'. The following passages, which I quote selectively, are taken from the resolution passed by the participants.

> Gene and reproductive technologies are the most recent attempt of the alliance of interests between business, science, politics and the military to open up new world markets. This is done to solve the ever-growing marketing problems faced by capital (surplus production, falling profit margins, limits to growth) in seeking new outlets. The new 'territories' now being conquered, dissected, invaded, appropriated and industrially used to make profits are plants, animal and human life, which are being subjugated to total control.
>
> It is especially the female body, with its unique capacity to create human life, which is being expropriated and dissected as raw material for the industrial production of humans. For us women, for nature and for the exploited people of this world this development is paramount to a declaration of war. For us women it means a further step towards the end of self-determination over our bodies, our ability to procreate and consequently our ultimate dependence on medical experts. We declare that we do not need or want this technology and that we fight it for what it is: a declaration of war on women and nature.[1]

This is strong language, with few hints of compromise or conditional acceptance. It is unmistakably anti-capitalist. The hostility to a certain kind of mechanistic science is clear. What comes across to me most strongly of all is a concern with women's loss of autonomy through the further development of medical science and technology. This is not just a piecemeal rejection of particular ideas or projects, but a vigorous and systematic attack on the whole scientific and medical programme of 'the new genetics', along with its philosophy of social engineering and its commercial linkages. It is an expression of rage about a whole complex of scientific, medical, administrative and industrial plans, seen as threatening women's lives around the world. This kind of criticism has been most powerfully voiced in West Germany, as part of the cultural movement that has given political strength and leverage to the Green Party. But similar sentiments have been expressed by radical feminists in a number of Western and industrialising countries. It has led, for example, to an international network of dissenting feminist groups, hostile to reproductive technology, called FINNRAGE. Although their primary concern is with women's oppression, statements such as these contain some of the most trenchantly critical comments about capitalism as a global economic system to be found in contemporary political debate.

Statements like the one above engage many, but not all, of my political sympathies. I believe that we live in a time of global economic change that is likely to perpetuate an unjust economic order. I share the view that a whole range of sectional interests will have to be swept aside to carry through these strategic changes. I would accept that such restructuring takes for granted a profoundly destructive attitude to nature, and a widespread lack of concern with alienation and economic exploitation at work. Furthermore, the regeneration of the world economy, with the breaking of the power of organised labour and the relocation of production, has profound implications for levels of employment, the nature of work, the time and freedom available for leisure and consumption, and the way in which households are run. Many of these things are likely to strike directly at women's incomes, independence of action and social status. Moreover, fundamental to this process is a re-definition of needs and here too women's lives will be profoundly affected. Sadly it is all too likely that many new products and services will not meet their needs, as they would prefer them to be met, and that some will expose them to new risks.

Nonetheless, it is hard to see the kind of resistance and opposition described above as a viable political programme. Its strength lies in its value to generate reflection and to force a debate about the values being built into new technologies. But to organise around such a radical renunciation and use it practically would be unfeasible, not least because it entails such a sweeping rejection of research and technology, which many people still value, despite the constraints they impose. Not everyone has the strength of purpose to resist a compromise.

From the other end of the political spectrum new procedures like ivf have come in for equally vigorous denunciation, but for very different

178

reasons. Again, however, the new genetics is seen as unnatural. In particular the human embryo has become a symbol of vulnerability and our concern for its supposed rights a token of our awareness of what makes us human. Yet the fundamental reference is not to women's rights and freedoms but to intuitions and beliefs about human identity. This passage comes from the *Salisbury Review*.

> There are three essentialy 'biological' propositions which, in the light of human history, to make no higher claim for them, must command respect.
> (1) That the conception of the child apart from normal intercourse, if widely practised, as a result of facilitating legislation, will ultimately produce a large and increasing category of children in whom the sense of identity and 'personhood' has been weakened and destroyed. A child's sense of identity derives primarily from its knowledge that 'X' is its father and 'Y' its mother.
> (2) That the process will weaken and gradually destroy the unique biological 'blood' link between parent and child.
> (3) That the process will weaken and gradually destroy the foundations of the almost-biological unity between husband and wife which is intimately bound up with the knowledge that the child, the issue of their relationship, is, in a sense, the product of their own two bodies, minds, beings.
>
> In my view, these are the biological and psychological bases on which both marriage and the family depend, far more than they do on the prescriptions of either civil or religious law.[2]

This kind of question-begging, mystical silliness is intolerable. I cite it here to convey both the tone and the terms of another form of dissent, masquerading as some sort of folk wisdom rendered as propositions. Of course a sense of identity and belonging does matter, but it does not have to rest on 'blood links' (which I always thought was a type of sausage), and 'the almost-biological unity between husband and wife'.

Although B.A. Santamaria, whose words I have just quoted, does not speak of nature, others with very similar views have done so.

> The foetus may not make 'choices' but it responds to noises, even the mother's moods, and moves into comfortable positions very early on: it is already progressing through complex stages of becoming. It may not be possible to describe 'it' as a person, but it is not merely a 'physical basis': it is a living creature of promise . . . The way in which we behave towards such a being will indicate the degree to which we are capable of developing a whole adequate sense of man's relationship with the natural world. This is more important than our attitudes to, say, the extinction and preservation of species, the preservation of the atmosphere, the avoidance of nuclear pollution, the control of war, and the contest against disease; of

179

course, the need for us to find a new understanding of man's relationship with nature underlies all these. But it underlies especially our attitude to our own foetuses. And this issue affects what we think of ourselves: all our moral attitudes.[3]

Nature, it seems, has many uses. The fact of variety might tempt us to conclude that all references to some natural order or set of rules are equally worthless. This would be wrong. As I argued in the first chapter, conceptions of nature cannot bear the psychological, moral or legal weight that sometimes we would like them to bear. They are not and cannot be some ultimate guarantee of rectitude and good sense. Instead they codify the received wisdom. They express the feeling that particular rules and meanings can be justified. Accordingly, there is no obligation for us to try and purify our speech of these natural references, providing we recognise them for what they are. Moreover, the attempt would be bound to fail, since what all these writers correctly perceive is that we cannot do science without some underlying framework that guides our hand. Our ideas about what the natural world contains influence the projects that we pursue.

What role can moral philosophy play?

The Warnock report has appeared twice, once from Her Majesty's Stationery Office, bound in blue paper, once with Lady Warnock's name on the cover, and in a commercial paperback edition.[4] I greatly preferred the second edition, not only because it was cheaper, but also because it contained two essays on the role of philosophy in social decision-making, or, to put it more bluntly, in governing. In them Lady Warnock discusses, firstly, what morality means, and, secondly, how it relates to law-making.

> By themselves, then, neither utilitarianism nor a blind obedience to rules could solve the moral dilemmas the Inquiry was faced with. We were bound to have recourse to moral sentiment, to try, that is, to sort out what our feelings were, and to justify them. For that a decision is based on sentiment by no means entails that arguments cannot be adduced to support it. Nor are utilitarian arguments, based on possible benefits and harms ruled out. It is only that they will not suffice alone. What is essential is to recognise that sentiment has some part, and indeed a crucial part, in arriving at moral decisions. For if this is recognised, it may be less surprising that agreement is not always possible in matters of morality. We know that people's feelings differ. Therefore moral conflict may be unavoidable.[5]

She then goes on to discuss the problems of moving from moral discussions, where there is very likely to be considerable unresolved disagreement, to the formulation and application of the law, where the operational principles must be precise and generally agreed. Here the question has to be recast, to ask not only whether some action will be generally thought objectionable, but also to consider whether some people will think legal intervention to stop it a more serious moral wrong. In effect on every issue there are now two questions to consider, not just one. Thus in the case of surrogacy, the Committee were almost unanimous in their disapproval of the basic idea, but some members very reluctant to recommend legal action against it, because the creation of enforceable laws to stop it would require an unacceptable intrusion into everyone's personal lives.

Now I hold no particular brief for the Warnock report, although at various points I have endorsed some of its recommendations. I mention it not to praise or denigrate it, but because it illustrates what abstract reflection can do. The task that is constantly before us is to sort our tangled moral sentiments into some kind of consensual framework. Moral philosophy offers us the means by which to assess the force of arguments for and against particular technologies and procedures. One of my problems with the different responses to the new genetics discussed in the second section of this chapter is that they have a sort of take-it-or-leave-it quality; they pre-empt, rather than encourage, discussion. Indeed, although it saddens me to say it, the conservative article mentioned in fact offers slightly more purchase than the one from the Left.

Nonetheless, philosophy has two degenerate forms, into which it is quite easy to decay. One is unremitting abstraction; the other is an indiscriminate attack on all positions so that one is left without the basis for any convictions at all. The first is a form of obscurity, the second a recipe for paralysis. In assessing the sentiments aroused by reproductive technologies, I have tried to avoid both. I have aimed to steer clear of these difficulties, firstly, by making considerable use of analogy and, secondly, by documenting as fully as possible how people actually behave or say they feel in the situations where some use is being made of 'the new genetics' or in analogous contexts. A great deal of moral reasoning proceeds by supposition and unacknowledged generalisation from the merest handful of data. I have deliberately tried to find, display and use relevant empirical studies, since what people actually want and find tolerable is not always what we fondly imagine or suppose. Obviously at some point the factual reporting has

181

to stop but the social process of evaluating the new genetics could profit from a great deal more empirical investigation. Indeed the two aspects, sociological and moral/political, should interact with and reinforce one another. Too many of the empirical studies I mentioned took the technology more or less for granted; too few of the moral discussions were backed up with facts and figures. Nonetheless I was surprised, and I hope my readers are too, by the volume of data already collected.

Confidence and autonomy

Now let me return to my primary theme, the issue of confidence to act according to one's own sense of what is right. Although I have affirmed this principle at several places in the book – in relation to artificial insemination, informal non-commercial surrogacy and antenatal diagnosis particularly – it will also have struck some people that I do not endorse everything that people might want to do. The best way to get at what I mean is through a discussion of autonomy, which is very much a linked idea. If people are to act autonomously, not subject to the prejudices and controls of expert advisers, they must have confidence both in their own knowledge and their own moral judgment. Equally, the whole point of building people's confidence in their self-understanding and ability to live with particular decisions is to allow them to live autonomously and with respect for others. Remove either quality and people are constrained.

Howard Brody has had some interesting things to say about autonomy, as an idea with a history.[6] In medical ethics fifteen years ago, when the subject was first finding its professional feet, philosophers used to speak of autonomy rather baldly as a very straightforward and desirable thing which physicians had tended to overlook after centuries of paternalism. Since one would imagine that everyone would want to be treated as a sensible and mature individual perfectly capable of making choices for themselves, the task of the physician, so philosophers implied, was simply to offer the options and then to back off, lest the patient's autonomy be compromised.

Not surprisingly some doctors hit back, arguing that these ideas parodied their professional behaviour and their relations with people seeking treatment and advice. What has emerged from the ensuing discussions is something more subtle than the original all-or-nothing view of autonomy. People, we now perceive, are not so simply controlled by the professionals with whom they deal; nor is their autonomy necessarily enhanced by expert reticence. You do not always free someone by not getting involved with their decisions.

One more recent view of autonomy, discussed by Brody, takes these objections on board. It has four aspects: free action, authenticity, effective deliberation and moral reflection. Free action means knowing what one is doing and choosing voluntarily to do it. Amongst other things people acting freely in this sense may be under great emotional stress; they are not free of all pressure, but aware of decisions before them and able to handle them despite the pressure. Authenticity means acting in character, or in accord with important, pre-existing, deep-seated beliefs.

Effective deliberation refers to the rational weighing of the pros and cons of the various options, and deciding accordingly. This is often what is in people's minds when they talk of informed consent, acceptance based on enough information about different options and their consequences. At this point we may be starting to idealise the way in which people take decisions, but that is no reason to exclude this from a discussion of autonomy. The final aspect is moral reflection, where someone reviews their basic values and beliefs and assures themselves, if they can, that what they have decided is consistent with them. This too inevitably idealises how people operate but it is still quite reasonable to take this as something to be worked toward. It certainly imposes new and burdensome obligations on doctors and those seeking their advice. It also broadens out the question of autonomy into something to be developed and discussed in other situations and with other people. It is, then, not just a question of how people are treated by their doctors, but how as a society we empower people to handle distressing, complex and ambiguous medical decisions, whether we anticipate a general need for the confidence to act freely, authentically and effectively in such situations.

The truth, alas, is that as a society we do not take these things very seriously, partly because we avoid thinking about them, unless medical paternalism becomes intolerable, and partly because technology and medical expertise gets in the way. That is, we tend to focus on the professionals' dilemmas and what to do with technology once it has been created. Instead we could make personal responsibility and effective deliberation the primary issue, and ask how professional behaviour and the use of particular technologies will affect them. This stands much of medical ethics on its head. It is not that professional practice can be forgotten or allowed to take care of itself. What little moral discussion takes place within professions is still very useful. But it should not be the principal focus of our attention.

I think this goes some of the way to mitigating the inconsistencies in my use of the idea of autonomy, as something generally to be encouraged, but which at the same time must apparently be denied people if they go 'too far'. After all why should not one confidently and autonomously avail oneself of the services of a surrogate mother on the books of a commercial agency or freely choose the abortion of a female foetus? I could imagine someone taking either of these decisions in a way that would satisfy all of the four criteria listed above. Or to put it another way, why is Lesbian parenthood not going 'too far' in this book, whereas hiring a woman to bear a child apparently is?

To do justice to this question I would have to write another book, but basically my argument would be, harking back to what I said in Chapter 5, that with surrogacy autonomy is actually purchased by the commissioning couple. The very terms of the arrangement are inconsistent with the ideals on which our whole discussion of autonomy and confidence is based. What is created specifically to allow the transaction to go through is a distorted and shackled form of deliberation that keeps returning the same answer. As far as self-insemination by a Lesbian woman is concerned, her actions affect herself, those with whom she shares her life, and, most importantly, the people she will bring into the world. Contrary to what stereotype and prejudice may suggest, this is a setting in which people's confidence and self-image, including that of the resulting children, can be enhanced. Of course things could go wrong, but there is not the fundamental denial of someone else's autonomy that is essential to commercial surrogacy. I would make a very similar argument about sex predetermination. This too might seem a form of reproductive freedom we should allow people, but again in a society stratified by gender this option would strike directly at women's self-esteem.

This question leads us finally back to nature.

Nature or technology?

The commonest way in which to speak critically of technology is to contrast it with nature. The two work very effectively as opposites. When radical feminists write of 'gene technologies' as a declaration of war against nature the symbolism has real power. Even though in other contexts natural events and processes may be evoked very

184

differently, no one would be taken aback by this critical juxtaposition. Technology is alien and uncontrollable, a means to domination; nature is beneficent and organised, something to be understood but not mastered.

This critical manoeuvre is most obvious in the movements pressing for 'natural childbirth' and 'natural death'. In both cases what is being opposed is the tacit assumption by doctors that people need, and in their own interests will always submit to, technologically intensive assistance. In the case of childbirth women are asking more and more often for a greater degree of choice and control over how they give birth. The place of birth, the use of monitoring equipment, the delivery position, the use of anaesthetics and drugs, the number of people present and the nature and timing of surgical intervention are all being questioned as factors that unnecessarily complicate a natural process. Women are less and less convinced that such issues must be dictated by the doctors concerned. For them technology subtracts from their autonomy, and impedes their control over what is happening.

In the case of terminal illness and death more people, particularly in the US, are making prior arrangements to limit the medication and surgery they will receive, in order to spend their last days in a conscious state, to die with dignity and to control the hospital bills. Statutory provision for such action is referred to as the 'Natural Death Act'.

In both cases some of what high-technology medicine has to offer is being rejected quite decisively. In both cases what is being reclaimed is autonomy and choice, in place of the inertness created supposedly to facilitate a safe delivery or to prolong life for a few more days. Particularly in the case of childbirth women are seeking to alter the experience of giving birth, not only to feel more in control of events and decisions, but also to enter more fully into the experience. Indeed one of the striking things about the 'natural childbirth movement' is the emphasis on the experience of birth, which is rehearsed, imagined, savoured and remembered, as a major event in one's life, with a particular value. Death obviously works on our imaginations in different ways, but here too some braver spirits clearly feel that they want to leave this world with their faculties in order and without the undignified pretence that another operation or yet more drugs can alter what was always going to happen. They believe that they will be able to handle the experience, without needless technological intervention.

These examples are well worth pondering. Firstly, these alternative approaches to events at the poles of existence are often bitterly resisted by medical people, at all levels, not only in public debate, but more insidiously in hospital practice. This is very wrong, both because the lay organisations are very well informed and quite able to tackle all the technicalities and other questions, and because the major occupational hazard in any profession is complacency, as routine and expedient supposition block actual thought. Arguing the merits of nature against technology is a serious exercise, but not something that has to be left to professionals alone.

Secondly, this contrast works only with certain limits. Much hangs upon what one means by nature. It is the intentions behind the labels that are more important than their exact denotation. What women experience in a home birth, without electronic monitors and without the pressure and disruption of hospital shift-working is certainly different. It may feel more natural, but it is still very much culture-bound, and it is still very likely to depend on some technology, like a central heating system or plastic sheets. In other societies different procedures in different settings could also be called a natural birth with equal justification. Similarly a 'natural death' in an industrialised country is made easier by pain relief and by good food.

Thirdly, arranging things this way, and setting limits on how technology is used, can build people's confidence in their own powers. Their involvement with medicine is not alienating and depersonalising. They learn from the experience. It is not so much that 'technology' is absent, but reliance upon it is very different and much less constraining, and the whole procedure is organised and planned with this in mind.

This then would be the redemption of the 'new genetics', if it could become natural in this sense, if it could empower people to understand more of what happens within their bodies and to feel able to make serious reproductive choices, with support and advice, but on terms that are freely chosen, rather than dictated by the technology.

Considered from this angle much that I have been discussing in this book comes off rather badly, and we have major difficulties ahead of us to prevent things getting worse. Artificial or self-insemination is perhaps the most promising area, if we can face up to the difficult problems of secrecy. Here we could either institutionalise a very high degree of technical and moral supervision, which is the French approach, or we could devise ways of placing more and more power in the hands of those who want to conduct insemination for themselves, without abandoning all concern for technical standards. I very much favour the latter.

186

With ivf we are already into technology-based reproduction. At the end of Chapter 4 I suggested that the very drama involved might help women and men to come to terms more easily with infertility, if ivf failed. It is much more plausible to argue that with appropriate counselling people would come to see much of the heroics, expense and clinical experimentation as something of a waste of time. Morever, if we put more effort and money into helping people understand their bodies and their sexuality and feel confident to ask for more help and advice when they sense that some problem may threaten their fertility, then in the longer term we would do much more good. The problem is that this can only work over the longer term. In the meantime ivf programmes will continue to attract interest. They could be helped by a lot of lay consideration as to how they are organised. In particular we should concentrate on trying to ensure that people's confidence in themselves is enhanced, that they are not treated simply as captive research subjects, that they are fully informed about all the details of the procedure, and that people's feelings about the experience are fed back into the programme.

With surrogacy there is little hope. I could just about imagine that people embarking on it informally, as the woman carrying the child, or the receiving person/couple, could emerge from the experience feeling better about themselves and with greater assurance and self-awareness. Philip Parker argued something like this of some people in commercial surrogacy, which is pretty implausible. In any case I have already argued strongly against surrogacy on exactly these grounds. Much the same has to be said about sex predetermination.

Genetic intervention is something of a mixture. Genetic screening could be done in ways that help people feel more knowledgeable about their bodies and reproduction and more able to handle the decisions involved. That would mean that it would have to be taken much more seriously, supplemented with a lot of counselling and discussion, and taken out of the medical setting. Antenatal diagnosis is almost inevitably going to remain specialist advice, heavily oriented to hospital obstetric care. The most hopeful development would be a much greater recognition of the complexity of the experience and much more respect for the autonomy of the people centrally involved. Gene therapy should be approached in the same way; genetic engineering, on the other hand, fails my basic test of confidence-building. It assumes a degree of anxiety and concern on the part of would-be parents that ought to be handled in other ways.

The balance sheet, then, shows far more entries on the technology page, and very little that will help us develop a culture where medicine is organised to give people an understanding of their bodies and the power and confidence to take their own decisions about what should happen to them. In that sense much that we have discussed in this book remains unnatural and must be tackled as such.

Notes

Chapter 2 Artificial insemination: out of the night

1   P. Snowden and G. D. Mitchell, *The Artificial Family: A consideration of artificial insemination by donor* (London: George Allen & Unwin, 1981).

2   J. Kremer, B. W. Frijling and J. L. M. Nass, 'Psychosocial aspects of parenthood by artificial insemination donor', *The Lancet* (17 March 1984), 628.

3   R. G. Bunge, W. C. Keetel and J. K. Sherman, 'Clinical use of frozen semen: report of 4 cases', *Fertility and Sterility*, 5 (1954), 520.

4   One estimate in 1983 was of 2,500 live births after AID in the UK: see S. Teper and E. M. Symonds, 'Artificial insemination by donor: problems and perspectives' in *Proceedings of the Eugenics Society Symposium*, 19 (1983), 163-76. Figures for adoptions in the UK are given in *Social Trends* and the publications of the Office for Population Censuses and Surveys.

5   F. N. L. Poynter, 'Hunter, Spallanzani and the history of artificial insemination' in L. Stevens and R. Multhauf (eds), *Medicine, Science and Culture* (Baltimore: Johns Hopkins Press, 1968), 97- 113.

6   A. McLaren, *Reproductive Rituals: the perception of fertility in England from the sixteenth century to the nineteenth century* (London: Methuen, 1984).

7   E. Home, 'An account of the dissection of a hermaphrodite dog' *Philosophical Transactions of the Royal Society of London*, 89 (1799), 159-78.

8   A. T. Gregoire and R. C. Mayer, 'The impregnators', *Fertility and Sterility*, 16 (1965), 130-4.

9   A. A. Schorohowa, 'La fécondation artificielle dans l'espèce humaine', *Gynécologie et Obstétrique*, 15 (1927), 132-9.

10 M. H. Jackson, 'A medical service for the treatment of involuntary sterility', *Eugenics Review*, 36 (1945), 117-25; M. H. Jackson, 'Artificial insemination (donor)', *Eugenics Review*, 48 (1957), 203-5.

11 M. Barton, K. Walker and B. P. Wiesner, 'Artificial insemination', *British Medical Journal*, i (13 January 1945), 40-3.

12 *Report of the Departmental Committee on Human Artificial Insemination* (London: HMSO, 1960; Cmnd 1105) (Feversham report).

13 'Report of Panel on Human Artificial Insemination', *British Medical Journal Supplement* (7 April 1973), Appendix V, 3-5 (Peel report). For details of the first NHS clinic established see R. S. Ledward et al., 'The

establishment of a programme of AID semen within the NHS', *British Journal of Obstetrics and Gynaecology*, 82 (1976), 917-20.

14 R. Rowland, 'Community attitudes to artificial insemination by husband or donor, in vitro fertilisation and adoption', *Clinical Reproduction and Fertility* 2 (1983), 195-206.

15 *Report of the Committee of Inquiry into Human Fertilisation and Embryology* (London: HMSO, 1984; Cmnd 9314) (Warnock report): reprinted with an introduction, as M. Warnock, *A Question of Life* (Oxford: Blackwell, 1985).

16 M. Curie-Cohen, L. Luttrell and S. Shapiro, 'Current practice of artificial insemination by donor in the United States', *New England Journal of Medicine*, 300 (15 March 1979), 585-90.

17 G. David and J. Lansac, 'The organisation of the Center for the Study and the Preservation of Semen in France' in G. David and W. S. Price (eds), *Human Artificial Insemination and Semen Preservation* (New York: Plenum Press, 1980), 15-25.

18 These figures are from S. B. Novaes, 'Social integration of technical innovation: sperm banking and AID in France and the United States', *Social Science Information*, 24 (1985), 569-84.

19 R. D. Klein, 'Doing it ourselves: self-insemination' in R. Arditti, R. D. Klein and S. Minden (eds), *Test-Tube Women: What future for motherhood?* (London: Pandora, 1984), 382-90; see also F. Hornstein, 'Children by donor insemination: a new choice for lesbians', ibid., 373-81.

20 C. Strong and J. S. Schinfeld, 'The single woman and artificial insemination by donor', *Journal of Reproductive Medicine*, 29 (May 1984), 293-9; see also M. McGuire and N. J. Alexander, 'Artificial insemination of single women', *Fertility and Sterility*, 43 (1985), 182-4; and G. T. Perkoff, 'Artificial insemination in a Lesbian: a case analysis', *Archives of Internal Medicine*, 145 (1985), 527-31.

21 M. Kirkpatrick, C. Smith and R. Roy, 'Lesbian mothers and their children', *American Journal of Orthopsychiatry* 51 (July 1981), 545-51; K. G. Lewis, 'Children of lesbians; their point of view', *Social Work*, 25 (1980), 198-203; R. Green, 'Sexual identity of 37 children raised by homosexual or transexual parents', *American Journal of Psychiatry*, 135 (June 1978), 692-6; M. E. Hotvedt and J. B. Mandel, 'Children of lesbian mothers' in W. Paul, J. D. Weinrich and J. C. Gonsiorek (eds), *Homosexuality: social, psychological and biological issues* (Beverly Hills: Sage, 1982), 275-85; E. Lewin and T. A. Lyons, 'Everything in its place: the coexistence of lesbians and motherhood', ibid., 249-73; S. Golombok, A. Spencer and M. Rutter, 'Children in lesbian and single-parent households: psychosexual and psychiatric appraisal', *Journal of Child Psychology and Psychiatry*, 24 (1983), 551-72.

22 G. J. Stewart et al., 'Transmission of human T-cell lymphotrophic virus

Type III (HTLV-III) by artificial insemination by donor', *The Lancet* (14 September 1985), 581-4.

23 An article on the Nottingham-based AID clinic reported that several people in their programme had achieved extramarital pregnancies whilst attending the clinic, and that several others had been advised to do so by their doctor: R. S. Ledward et al., 'Social and environmental factors as criteria for success in artificial insemination by donor (AID)', *Journal of Biosocial Science* 14 (1982), 263-75.

24 E. Haimes and N. Timms, *Adoption, Identity and Social Policy: the search for distant relatives* (Aldershot: Gower, 1985).

25 E. Alder, 'Psychological aspects of AID' in A. Emery and I. Pullen (eds), *Psychological Aspects of Genetic Counselling* (London: Academic Press, 1984), 187-99.

26 Snowden and Mitchell, op. cit. (See note 1 above).

27 R. Rowland, 'Attitudes and opinions of donors on an artificial insemination by donor (AID) programme', *Clinical Reproduction and Fertility*, 2 (1983), 249-59; see also D. J. Handelsmann et al., 'Psychological and attitudinal profiles in donors for artificial insemination', *Fertility and Sterility*, 43 (1985), 95-101.

28 J. Teichman, *Illegitimacy: A philosophical examination* (Oxford: Blackwell, 1982).

29 Law Commission, *Family Law: Illegitimacy* (London: HMSO, 1982; Law Com. No. 118).

30 C. Holden, 'Sperm banks multiply as vasectomies gain popularity, *Science*, 176 (1972), 32.

31 I. Davies, 'Cryobanking and AIH', *Medico-Legal Journal*, 52 (1984) 242-7.

32 W. J. Broad, 'A bank for Nobel sperm', *Science*, 207 (21 March 1980) 1326-7; see also E. A. Carlson, *Genes, Radiation and Society: The life and work of H. J. Muller* (London: Cornell University Press, 1981). M. Wallace, 'Waiting for a sperm-bank genius', *Sunday Times* (4 July 1982), 13.

Chapter 3 In vitro fertilisation: experience and technique

1 For more detail than I have given here see M. H. Johnson and B. J. Everitt, *Essential Reproduction* (Oxford: Blackwell, 2nd edn., 1984); also R. G. Edwards, *Conception in the Human Female* (London: Academic Press, 1980).

2 N. Pfeffer and A. Woollett, *The Experience of Infertility*, (London: Virago, 1983).

3 P. P. Mahlstedt, 'Psychological components of infertility', *Fertility and Sterility*, 43 (1985), 335–46.

4 B. E. Menning, 'The emotional needs of infertile couples', *Fertility and Sterility*, 34 (1980), 313–19.

5 E. W. Freeman et al., 'Psychological evaluation and support in a program of in vitro fertilisation and embryo transfer, *Fertility and Sterility*, 43 (1985), 48–53.

6 Some work of this kind has already been done by Helen B. Holmes; she now plans to do more in New Zealand.

7 D. Greenfield et al., 'The role of the social worker in the in vitro fertilization program', *Social Work in Health Care*, 10 (Winter 1984), 71–79.

8 This is drawn from C. Wood and A. Westmore, *Test-Tube Conception* (London: Allen & Unwin, 1984) and reflects the Australian procedures.

9 E. Alder, talk on her research on people attending an AID clinic and the ivf program in Edinburgh, Marcé society meeting, London, July 1985.

10 Strictly speaking the whole procedure involved is in vitro fertilisation and embryo transfer, and this is sometimes abbreviated to ivf and et. The power of Hollywood being what it is there are connotations to the latter that are best avoided. I shall use ivf to mean both, unless the context requires me to be more specific.

11 Alder, op. cit. (see note 9 above).

12 C. Wood et al., 'Clinical implications of developments in in vitro fertilisation', *British Medical Journal*, 289 (1984), 978–980.

13 E. Alder and A. A. Templeton, 'Patient reaction to ivf treatment', *Lancet*, i (1985), 168.

14 M. R. Soules, 'The in vitro fertilisation pregnancy rate: let's be honest with one another', *Fertility and Sterility*, 43 (1985), 511–12.

15 A. Leathard, *District Health Authority Family Planning Services in England and Wales* (London: Family Planning Association, 1985).

16 National Association for the Childless, 318, Summer Lane, Birmingham B19 3RL; CHILD, 'Farthings', Gaunts Road, Pawlett, Nr Bridgwater, Somerset; British Organisation of Non-Parents, BM Box 5866, London WC1N 3XX. Other addresses are in N. Pfeffer and A. Woollett, *The Experience of Infertility*, (London: Virago, 1983).

17 See the issue for 1 December 1980 of the *American Journal of Obstetrics and Gynecology* which is entirely devoted to pelvic inflammatory disease; also 'Infertility and sexually transmitted disease: a public health challenge', *Population Reports*, Series L No. 4 (July 1983); S. O. Aral et al., 'Self-reported pelvic inflammatory disease in the US: a common occurrence', *American Journal of Public Health*, 75 (1985), 1216–18.

Chapter 4 In vitro fertilisation: from animal agriculture to medical venture capital

1 A. Westmore, 'History' in C. Wood and A. Trounson, *Clinical In Vitro*

Fertilization (Berlin: Springer, 1984), 1–10; see also G. Sarton, 'The discovery of the mammalian egg and the foundation of modern embryology', *Isis*, 16 (1931), 315–30; C. W. Bodemer, 'The biology of the blastocyst in historical perspective', in R. J. Blandau (ed.), *The Biology of the Blastocyst* (London: University of Chicago Press, 1971); K. J. Betteridge, 'An historical look at embryo transfer', *Journal of Reproduction and Fertility*, 62 (1981), 1–13.

2   F. H. A. Marshall, 'Walter Heape, 1855–1929', *Proceedings of the Royal Society*, B. 106 (1930), xv–xviii.

3   W. Heape, 'Preliminary note on the transplantation and growth of mammalian ova within a uterine foster-mother', *Proceedings of the Royal Society*, 48 (1891), 457–8; idem, 'Further note on . . .', ibid., 62 (1897–8), 178–83.

4   M. Borrell, 'Organotherapy and reproductive physiology', *Journal of the History of Biology*, 18 (1985), 1–30.

5   E. Allen and E. A. Doisy, 'An ovarian hormone: preliminary report of its localisation, extraction and partial purification', *Journal of the American Medical Association*, 81 (1923), 819–21.

6   E. Allen et al., 'Recovery of human ova from the uterine tubes: time of ovulation in the menstrual cycle', *Journal of the American Medical Association*, 91 (1928), 1018–20.

7   B. Brackett, 'Recent progress in investigations of fertilisation in vitro' in R. J. Blandau, op. cit. (see note 1 above), 329–48.

8   Anon., 'Conception in a watch glass', *New England Journal of Medicine*, 217 (1937), 678.

9   L. McLoughlin, *The Pill, John Rock and the Church: The biography of a revolution* (Boston: Little, Brown, 1982).

10 A. T. Hertig et al., 'Thirty four fertilised human ova, good, bad and indifferent, recovered from 210 women of known fertility: a study of biologic wastage in early human pregnancy', *Paediatrics*, 23 (1959), 202–11.

11 J. Rock and M. F. Menkin, 'In vitro fertilisation and cleavage of human ovarian eggs', *Science*, 100 (1944), 105–7.

12 M. C. Chang, 'Fertilisation of rabbit ova in vitro', *Nature*, 184 (1959), 466–7; for a sceptical contemporary review of the evidence up to this point see C. R. Austin and M. W. H. Bishop, 'Ferilisation in mammals', *Biological Reviews*, 32 (1957), 296–349.

13 M. J. Kottler, 'From 48 to 46: cytological technique, preconception and the counting of human chromosomes', *Bulletin of the History of Medicine*, 48 (1974), 465–502.

14 R. G. Edwards, 'Maturation in vitro of human ovarian oocytes', *The Lancet*, ii (1965), 926–9.

15 R. G. Edwards, B. D. Bavister and P. C. Steptoe, 'Early stages of fertilisation in vitro of human oocytes matured in vitro', *Nature*, 221

(1969) 632–5; see also V. Rothschild, 'Did fertilisation occur?' ibid., 981.

16 R. G. Edwards, P. C. Steptoe and J. M. Purdy, 'Fertilisation and cleavage in vitro of preovulatory human oocytes', *Nature*, 227 (1970), 1307–10.

17 W. J. Sweeney and L. S. Goldsmith, 'Test tube babies: medical and legal considerations', *Journal of Legal Medicine*, 2 (1980), 1–16.

18 D. de Kretzer et al., 'Transfer of a human zygote', *The Lancet*, ii (1973), 728–9.

19 Anon., 'Embryo transplants', *British Medical Journal*, iii (1974), 238.

20 R. G. Edwards and P. Steptoe, *A Matter of Life: The story of a medical breakthrough* (London: Hutchinson, 1980).

21 C. Grobstein et al., 'External human fertilisation: an evaluation of policy', *Science*, 222 (1983), 127–33.

22 J. F. Henahan, 'Fertilisation, embryo transfer procedures raise many questions', *Journal of the American Medical Association*, 252 (1984), 127–33.

23 P. Soupart, 'Current status of in vitro fertilisation and embryo transfer in man [sic]', *Clinical Obstetrics and Gynecology*, 23 (1980), 683–717.

24 M. Clarke, 'Embryo research: Chances of legislation fade', *Nature*, 318 (21 November 1985), 197; M. Clarke, 'Ivf: another bill bites the dust', *Nature*, 319 (30 January 1986), 349.

25 R. Walgate, 'Human embryology: France seeks a policy in haste', *Nature*, 313 (28 February 1985), 728.

26 H. B. Holmes, 'And in the Netherlands: Guidelines for ivf', *Hastings Center Report* (August 1985), 6.

27 P. Singer, 'Making laws on making babies', *Hastings Center Report* (August 1985), 5–6.

28 *Creating Children: A uniform approach to the practice of reproductive technology in Australia* (Report of the Asche Committee on Reproductive Technology to the Family Law Council, 29 May 1985).

29 P. McIntosh and C. Miller, 'Uni council approves the sale of IVF technology', *The Age* (19 March 1985), 3; P. McIntosh, 'American to market our IVF technology overseas', *The Age* (2 June 1985), p. 5.

30 Harman letter to *The Age*, 22 March 1985, quoted in R. Koval, 'IVF sale raises issues for college academics', *Journal of Advanced Education* (May 1985), 7; see also R. Koval, 'Women, birth and power', *Australian Society* (July 1985), 6–8.

Chapter 5 Evaluating the criticisms of ivf

1 Criticism of natural law arguments of this kind is presented very well in P. Singer and D. Wells, *The Reproduction Revolution: New ways of making babies* (London: Oxford University Press, 1984).

2   See, for example, T. Iglesias, 'Social and ethical aspects of in vitro fertilisation' in *Test Tube Babies: A Christian view* (Oxford: Becket Publications, 1984), 75–104.

3   R. Barnes, 'The many perils that face test-tube babies', *Pulse* (3 June 1978), 1; H. O. Tiefel, 'Human in vitro fertilisation', *Journal of the American Medical Association*, 247 (1982), 3235–43.

4   See *A Matter of Life*, p. 111; see also R. G. Edwards, *Conception in the Human Female* (London: Academic Press, 1980), 1015–20.

5   J. D. Biggers, 'In vitro fertilisation and embryo transfer in human beings' *New England Journal of Medicine*, 304 (1981), 336–41.

6   J. J. Schlesselman, 'How does one assess the risks of chromosome abnormalities from human in vitro fertilisation', *American Journal of Obstetrics and Gynecology*, 135 (1979), 135–48.

7   J. F. Henahan, 'Fertilisation, embryo transfer procedures raise many questions', *Journal of the American Medical Association*, 252 (1984), 877–82.

8   Dr. V. Beral, Personal communication, 9 August 1985.

9   G. Corea, 'Reproductive technology and women', *Science for the People* (September/October 1980), 3–14; see also Barnes, op. cit. (note 3 above).

10 M. R. Soules, 'The in vitro fertilisation pregnancy rate: let's be honest with one another', *Fertility and Sterility* 43 (April 1985) 511–12; see also G. Corea and S. Ince, 'Ivf a game for losers at half US clinics', *Medical Tribune* (3 July 1985), 1.

11 C. Wood et al., 'Factors influencing pregnancy rates following ivf', *Fertility and Sterility*, 43 (1985), 245–54.

12 M. Seppala et al., 'The world collaborative report on in vitro fertilisation and embryo replacement: current state of the art in January 1984', in M. Seppala and R. G. Edwards (eds), *In Vitro Fertilisation and Embryo Transfer* (Annals of the New York Academy of Sciences, 442, 1985), 558–63.

13 M. Johnston et al., 'Expectations of success and stress associated with in vitro fertilisation and embryo transfer' (unpublished manuscript); see also E. W. Freeman et al., 'Psychological evaluation and support in a program of in vitro fertilisation and embryo transfer', *Fertility and Sterility*, 43 (1985), 48–53.

14 H. Leridon, *Human Fertility: The basic components* (London: University of Chicago Press, 1977); see also Schlesselman, op. cit. (see note 6 above).

15 See Soules, op. cit. (note 10 above).

16 S. Aral and W. Cates, 'The increasing concern with infertility: why now?', *Journal of the American Medical Association*, 250 (1983), 2327–31; M. P. Mollenkamp, 'The increasing concern with infertility', ibid., 252 (1984), 208.

17 E. Campbell, 'Becoming voluntarily childless: an exploratory study in a Scottish city', *Social Biology*, 30 (1983), 307–17; J. R. Wilkie, 'Involuntary childlessness in the United States', *Zeitschrift für Bevölkerswissenschaft*, 10 (1984), 37–52; D. L. Poston and K. B. Kramer, 'Voluntary and involuntary childlessness in the United States, 1955–1973', *Social Biology*, 30 (1983), 290–306.

18 W. D. Mosher and W. F. Pratt, 'Fecundity and infertility in the United States, 1965–82', *National Center for Health Statistics, Advance Data*, No. 104, (11 February 1985); W. D. Mosher, 'Reproductive impairments in the United States, 1965–82', *Demography* (forthcoming).

19 C. Grobstein, M. Flower and J. Mendeloff, 'External human fertilisation: an evaluation of policy', *Science*, 222 (1983), 127–33.

20 M. G. R. Hull et al., 'Population study of causes, treatment and outcome of infertility', *British Medical Journal*, 291 (14 December 1985), 1693–7.

21 Data from the DHSS.

22 The technical variations include surgical transfer of the ovum within the Fallopian tube, surgical removal of sperm from the epididymis, microinjection of sperm into ova, and embryo transfer to women without functional ovaries; see R. H. Asch., 'Pregnancy after translaparoscopic gamete intrafallopian transfer', *The Lancet* (3 November 1984), 1034–5; see also G. D. Hodgen, 'In vitro fertilisation and alternatives', *Journal of the American Medical Association*, 246 (1981), 590–7; G. D. Hodgen, 'Surrogate embryo transfer combined with estrogen-progesterone therapy in monkeys: implantation, gestation and delivery without ovaries', *Journal of the American Medical Association*, 250 (1983), 2167–71; P. Lutjen et al., 'Establishment and maintenance of pregnancy using in vitro fertilisation and embryo donation in a patient with a primary ovarian failure', *Nature*, 307 (1984), 174.

23 A. A. Templeton et al., 'The recovery of pre-ovulatory oocytes using a fixed schedule of ovulation induction and follicle aspiration', *British Journal of Obstetrics and Gynaecology*, 91 (1984), 148–54.

24 A. Bernard et al., 'Effects of cryoprotectants on human oocytes', *The Lancet* (16 March 1985), 632–3.

25 A. Trounson et al., 'Pregnancy established in an infertile patient after transfer of a donated embryo fertilised in vitro', *British Medical Journal*, 286 (1983), 835–8.

26 R. Scott, *The Body as Property* (New York: Viking Press, 1981); R. P. S. Jansen, 'Sperm and ova as property', *Journal of Medical Ethics*, 10 (1984), 32–7.

27 C. B. Coulam, 'Freezing embryos', *Fertility and Sterility*, 42 (1984), 184–6; D. T. Ozar, 'The case against thawing unused frozen embryos', *Hastings Center Report*, (August 1985), 7–12.

28 A. Trounson and L. Mohr, 'Human pregnancy following cryopreser-

vation thawing and transfer of an eight-cell embryo', *Nature*, 305 (1983), 707–9; G. H. Zeilmaker et al., 'Two pregnancies following transfer of intact frozen-thawed embryos', *Fertility and Sterility*, 42 (1984), 293; B. G. Downing et al., 'Birth after transfer of cryopreserved embryos', *Medical Journal of Australia*, 142 (1985), 409–10; Anon, 'Frozen embryo born', *The Guardian* (9 March 1985).

29 J. Sellar, 'Australian ivf: orphan embryos', *Nature*, 309 (1984), 938.

30 Report of the Advisory Committee to the DHSS, *The Use of Fetal Material for Research* (London: HMSO, 1972); NHMRC Report, 'Ethics in medical research involving the human fetus and human fetal tissue', *Medical Journal of Australia* (12 May 1984), 610–20.

31 O. Sattaur, 'New conception threatened by old morality', *New Scientist* (27 September 1984), 12–17; H. J. Evans and A. McLaren, 'Unborn Children (Protection) Act', *Nature*, 314 (1985), 127–8; for a different view see M. Rayner, 'Experiments on embryos: stick to the facts', *New Scientist* (27 February 1986), 54.

32 J. M. Goldenring, 'Development of the fetal brain', *New England Journal of Medicine*, 307 (1982), 564; T. Kushner, 'Having a life versus being alive', *Journal of Medical Ethics*, 10 (1984), 5–8; G. E. Jones, 'Fetal brain waves and personhood', *Journal of Medical Ethics,* 10 (1984), 216–8.

Chapter 6 Surrogacy

1 For general discussions of surrogacy see J. A. Robertson, 'Surrogate motherhood: not so novel after all', *Hastings Center Report*, 13 (1983), 28–34; G. J. Annas, 'Making babies without sex: the law and the profits', *American Journal of Public Health*, 74 (1984), 1415–17; G. J. Annas, S. Elias, 'In vitro fertilisation and embryo transfer: Medico-legal aspects of a new technique to create a family', *Family Law Quarterly*, 17 (1983), 199–223. M. Warnock, 'Legal surrogacy – not for love or money?' *The Listener*, 113 (1985), 2–4; Anon., 'Surrogacy falsely in the dock' [editorial], *Nature*, 313 (1985), 83; Anon., 'Is buying babies bad?' [editorial], *The Economist* (12 January 1985), 12; B. Cohen, 'Surrogate mothers: whose baby is it?' *American Journal of Law and Medicine*, 10 (Fall 84), 243–86.

2 See W. Greengross, D. Davies, 'Expression of dissent: A. Surrogacy' in *Report of the Committee of Enquiry into Human Fertilisation and Embryology* (London: HMSO, 1984), 87–9. In a conference in Oxford in July 1985 Lady Warnock said that this was her view too, but had felt unable to join the dissenters in the report, because of her position. Dr Robert Winston of the Hammersmith Hospital also declared himself in favour of surrogacy in very special cases on the BBC TV *Panorama* programme, 14 January 1985.

3   This was the case for the first client, a Lebanese man, of the American lawyer, Noel Keane, who now makes surrogacy arrangements as part of his legal practice; see N. P. Keane, 'Surrogate motherhood: past, present and future' in *Progress in Clinical and Biological Research*, 139 (1983), 155–64.

4   P. J. Parker, 'Motivation of surrogate mothers: initial findings', *American Journal of Psychiatry*, 140 (1983), 117–18.

5   'Surrogacy Arrangements Bill', *Hansard*, 77 (15 April 1985), 25–55; *Hansard*, 79 (13 May 1985), 115–22, continued in *Hansard*, 79 (14 May 1985), 123–33.

6   K. Clarke, Minister of Health, speaking in the House of Commons 3rd reading debate: see *Hansard*, 79 (14 May 1985), 130; see also K. Puttick, 'Surrogacy Arrangements Act', *New Law Journal*, 135 (1985), 849–50.

7   See 'The week in Plymouth', *British Medical Journal* (6 July 1985), 64–85.

8   D. Parker, 'Surrogate mothering: an overview', *Family Law*, 14 (1984), 140–3.

9   P. Toynbee, Interview with Dr R. Levin, *Guardian* (18 October 1982), 8.

10 J. R. S. Pritchard, 'A market for babies', *University of Toronto Law Journal*, 24 (1984), 341–57.

11 'A v C', *Family Law*, 8 (1978), 170–1

12 L. Davis, 'Surrogate parenting', *New Law Journal*, 134 (1984), 707–8; Annas, op cit. (see note 1 above) cites instances of very much higher figures.

13 S. Ince, 'Inside the surrogate industry' in R. Arditti, R. D. Klein and S. Minden (eds) *Test-Tube Women: what future for motherhood?* (London: Pandora, 1984), 99–116.

14 P. Toynbee, Interview with Kim Cotton, *Guardian* (1 July 1985), 9; Anon., 'Surrogate mothers', *British Medical Journal*, 290 (1985), 308; K. Cotton and D. Winn, *For Love and Money* (London: Dorling Kinnersley, 1985).

15 Dr R. Levin, see note 9 above.

16 Ibid.

17 Precisely this point is overlooked by E. Page in his article, 'Whose baby?', *The Times Higher Education Supplement*, (16 August 1985), 12.

18 I. Davies, 'Contracts to bear children', *Journal of Medical Ethics*, 11 (1985), 61–5.

19 M. Bustillo et al., 'Delivery of a healthy infant following non-surgical ovum transfer', *Journal of the American Medical Association*, 251 (1984), 889.

20 M. Bustillo et al., 'Non-surgical ovum transfer as a treatment in infertile women: preliminary experience', *Journal of the American Medical Association*, 251 (1984), 1171–73.

Chapter 7 Sex predetermination

1 A. Etzioni, 'Sex control, science and society', *Science*, 161 (1968), 1107–12; for a thorough historical and cross-cultural discussion of sex ratios and their effect on social mores see M. Guttentag and P. F. Secord, *Too Many Women? The sex ratio question* (London: Sage, 1983).

2 J. Hanmer and P. Allen, 'Reproductive engineering: the final solution?' in L. Birke et al., *Alice through the Microscope: The power of science over women's lives* (London: Virago, 1980), 208–27.

3 There are a number of very good reviews of the whole field; these include N. E. Williamson, *Sons or Daughters: A cross-cultural survey of parental preferences* (London: Sage, 1976); N. E. Williamson, 'Sex preferences, sex control and the status of women', *Signs*, 1 (1976), 847–62; B. B. Hoskins and H. B. Holmes, 'Technology and prenatal femicide', in R. Arditti, R. D. Klein and S. Minden (eds) *Test-Tube Women: What future for motherhood?* (London: Pandora, 1984), 237–55; H. B. Holmes, 'Sex preselection: eugenics for everyone?' in J. Humber and R. Almeder (eds), *Biomedical Ethics Reviews, 1985* (Clifton, N. J.: Humana Press, 1986), 39–71.

4 These complications include abnormalities in germ cells which lead to different chromosome combinations such as XXY and XYY, the developmental abnormality of mosaicism, where different sets of cells have different chromosome combinations and so the assignment of sex can be complicated, rare genetic conditions where individuals appear female at birth and become male at puberty, and the recently discovered cases of XX males, where one X-chromosome has picked up a small piece of Y-chromosome and therefore functions as one.

5 E. Croll, 'China's first-born nightmare returns', *Guardian* (28 October 1983); J. Mirsky, 'Return of the baby killers', *New Statesman*, 111 (21 March 1986), 19–20.

6 L. L. Cederqvist and F. Fuchs, 'Antenatal sex determination: a historical review', *Clinical Obstetrics and Gynecology*, 13 (1970), 159–77.

7 D. Rorvik and L. B. Shettles, *Choose Your Baby's Sex* (New York: Dodd, Mead, 1977).

8 R. Guerrero, 'Association of the type and time of insemination within the menstrual cycle and the human sex ratio at birth', *New England Journal of Medicine*, 6 (1974), 367–71; W. H. James, 'Cycle day of insemination, coital rate and sex ratio', *The Lancet* (16 January 1971).

9 B. Simcock, 'Sons and daughters – a sex preselection study', *Medical Journal of Australia*, 142 (1985), 541–2; see also J. T. France et al., 'A prospective study of the preselection of sex of offspring by timing intercourse relative to ovulation', *Fertility and Sterility* 41 (1984), 894–900.

10 Guttentag and Secord, op. cit. (see note 1 above).

11 J. Rakusen, 'Depo-Provera: the extent of the problem . . .' in H. Roberts (eds), *Women, Health and Reproduction* (London: Routledge & Kegan Paul, 1981), 75–108.

12 G. L. Gledhill, 'Control of mammalian sex ratio by sexing sperm', *Fertility and Sterility*, 40 (1983), 572–4.

13 S. L. Corson, 'Preconceptual female gender selection', *Fertility and Sterility*, 40 (1983), 384–5.

14 W. C. D. Hare and K. J. Betteridge, 'Animal sexing patents', *Nature*, 303 (1983), 654.

15 D. Shapley, 'Cattle breeding: techniques for sexing embryos now possible', *Nature*, 301 (1983), 101.

16 T. Beardsley, 'Embryo sexing: cattle now, people next?', *Nature*, 304 (1983), 301.

17 J. C. Hobbins, 'Determination of fetal sex in early pregnancy', *New England Journal of Medicine*, 309 (1983), 979–80.

18 For a fuller discussion of sex-linked diseases see A. M. Winchester and T. R. Mertens, *Human Genetics* (London: Merrill, 1983). My remarks in the text on the statistics of inheritance only apply to recessive conditions carried on the X-chromosome.

19 V. McKusick, 'The royal haemophilia', *Scientific American*, 213 (February 1965), 88–95.

20 P. Jones et al., 'AIDS and haemophilia: morbidity and mortality in a well-defined population', *British Medical Journal*, 291 (1985), 695–9.

21 G. G. Brownlee, 'Factor VIII cloned', *Nature*, 312 (1984), 307.

22 A. Ramanamma and U. Bambawale, 'The mania for sons', *Social Science and Medicine*, 14B (1980), 107–10; see also the criticisms in the same journal, in 1982, pp. 879–85.

23 V. Roggencamp, 'Abortion of a special kind: male sex selection in India' in R. Arditti et al., op. cit. (see note 3 above), 266–78.

24 R. Jeffery et al., 'Female infanticide and amniocentesis', *Social Science and Medicine*, 19 (1984), 1207–12.

25 C. F. Westoff and R. R. Rindfuss, 'Sex preselection in the United States', *Science*, 184 (1974), 633–6: also K. R. Widmer, 'Determining the impact of sex preferences on fertility: a demonstration study', *Demography,*, 18 (1981), 27–37.

26 'Foetal sex determination by sex chromatin prediction of chorionic villi cells during early pregnancy', *Chinese Medical Journal*, 1 (1975), 116–26.

27 A. R. Pebley and C. F. Westoff, 'Women's sex preferences in the United States: 1970 to 1975', *Demography*, 19 (1982), 177–89.

28 F. Gilroy and R. Steinbacher, 'Preselection of child's sex: technological utilization and feminism', *Psychological Reports*, 53 (1983), 671–6.

Chapter 8 Antenatal diagnosis: an inglorious past, a hi-tech future

1   D. J. Kevles, *In the Name of Eugenics: Genetics and the uses of human heredity* (New York: Knopf, 1985); D. A. Mackenzie, *Statistics in Britain,* 1865–1930 (Edinburgh: Edinburgh University Press, 1982).
2   C. K. Chan, 'Eugenics on the rise: a report from Singapore', *International Journal of Health Services,* 15 (1985), 707–12.
3   E. H. Ackerknecht, 'Diathesis: the word and the concept in medical history', *Bulletin of the History of Medicine,* 56 (1982), 317–25; in this century one of the first people to demonstrate the hereditary basis of certain diseases was Sir Archibald Garrod, in a series of famous papers on 'inborn errors of metabolism'; see A. G. Bearn and E. D. Miller, 'Archibald Garrod and the development of the concept of inborn errors of metabolism', *Bulletin of the History of Medicine,* 53 (1979), 315–28.
4   B. Childs, 'Persistent echoes of the nature-nurture argument', *American Journal of Human Genetics,* 29 (1977), 1–13.
5   E. J. Yoxen, 'Constructing genetic diseases' in P. Wright and A. Treacher (eds), *The Problem of Medical Knowledge* (Edinburgh: Edinburgh University Press, 1982), 144–61.
6   F. D. Ledley et al., 'Molecular biology of phenylalanine hydroxylase and phenylketonuria', *Trends in Genetics,* 1 (November, 1985), 309–13. PKU affects approximately 1 in 10,000 newborn infants in the US and Europe. One of the problems for women who are themselves phenyl- ketonuric arises if they want to have children. There may be 500 such women in the UK by the year 2,000. By adulthood they have come to tolerate the high blood levels of phenylalanine, but during pregnancy the foetus will be exposed to damagingly high levels unless they return to the phenylalanine-free diet. This might be a case where the use of a surrogate mother would make sense. It is also now possible to diagnose PKU antenatally, which would be of some use in reassuring people with PKU that their children are not affected; see B. Barwell, 'Phenylketonuria – a problem in pregnancy', *Health Visitor,* 55 (1982), 345–6.
7   R. Kenen, 'Genetic counselling: the development of a new inter- disciplinary occupational field', *Social Science and Medicine,* 18 (1984), 541–9.
8   P. Harper, 'Genetic counselling and prenatal diagnosis', *British Medical Bulletin,* 39 (1983), 302–9.
9   D. W. Goodner, 'Prenatal diagnosis: an historical perspective;, *Clinical Obstetrics and Gynecology,* 19 (1976), 837–40.
10 A. C. Turnbull and I. Z. Mackenzie, 'Second-trimester amniocentesis and the termination of pregnancy', *British Medical Bulletin,* 39 (1983), 315–21.
11 R. Faden et al., 'Pregnant women's attitudes toward the abortion of defective fetuses' *Population and Environment: Behavioral and Social Issues,* 6 (1983) 197–209.

12 T. Engelhardt, 'Current controversies in obstetrics: wrongful life and fetal surgery', *American Journal of Obstetrics and Gynecology*, 151 (1985), 313–21.

13 J. Lejuene, 'On the nature of men', *American Journal of Human Genetics*, 22 (1970), 121–8.

14 T. Winchester and H. Mertens, *Human Genetics*, 22 (London: Merrill, 1983). There is also the phenomenon of mosaicism, where some cells in the body have 47 chromosomes and some 46. This is thought to arise from a developmental abnormality in the early embryo. With Down's syndrome the effects are milder than with trisomy-21 in all cells.

15 M. Ferguson-Smith, 'Prenatal chromosome analysis and its impact on the birth incidence of chromosome disorders', *British Medical Bulletin*, 39 (1983), 355–64.

16 A. C. Turnbull and I. Z. Mackenzie, op. cit. (note 10 above).

17 M. Ferguson-Smith, 'The reduction of anencephalic and spina bifida births by maternal serum alphafetoprotein screening', *British Medical Bulletin*, 39 (1983), 365–72.

18 D. Dickson, 'Alpha-fetal protein screening: too hot to handle?' *Nature*, 280 (5 July 1979), 6–7.

19 G. J. Annas, 'Is a genetic screening test ready when the lawyers say it is?' *Hastings Center Report*, 15 (December 1985), 16–18.

20 K. Spencer and P. Carpenter, 'Screening for Down's syndrome using serum α-fetoprotein: a retrospective study indicating caution', *British Medical Journal*, 290 (1985), 1940–43.

21 B. K. Rothman, 'The products of conception: the social context of reproductive choices', *Journal of Medical Ethics*, 11 (1985), 188–92.

22 See, for example, S. C. Finley et al., 'Participants' reaction to amniocentesis and prenatal genetic studies', *Journal of the American Medical Association*, 238 (1977), 2377–9; N. Rice, R. Doherty, 'Reflections on prenatal diagnosis: the consumers' views', *Social Work in Health Care*, 8 (1982), 47–57; C Nielsen, 'An encounter with modern technology: women's experiences with amniocentesis', *Women and Health*, 6 (1981), 109–24; B. K. Burton et al., 'The psychological impact of false positive elevations of maternal serum a-fetal protein', *American Journal of Obstetrics and Gynecology*, 151 (1985), 77–82.

23 S. Bundey, 'Attitudes of 40-year-old college graduates towards amniocentesis', *British Medical Journal* (1978), 1475–7.

24 P. Donnai et al., 'Attitudes of patients after "genetic" termination of pregnancy', *British Medical Journal*, 282 (1981), 621–2.

25 W. Farrant, 'Who's for amniocentesis: the politics of prenatal screening', in H. Homans (ed.), *The Sexual Politics of Reproduction* (Aldershot: Gower, 1985), 96–122.

26 See, for example, the editorial, 'Human guinea pigs have a choice', *Nature*, 300 (1982), 565.

27 C. Faulder, *Whose Body Is It? The troubling issue of informed consent* (London: Virago, 1985).

28 The study is being organised by Prof. N. J. Wald of St Bartholomew's Hospital, London: see N. J. Wald and P. E. Polani, 'Neural-tube defects and vitamins: the need for a randomized clinical trial', *British Journal of Obstetrics and Gynaecology*, 91 (1984), 516–23.

29 D. J. Weatherall, 'Prenatal diagnosis of thalassaemia', *British Medical Journal*, 288 (1984), 1321–2.

30 L. R. Davis, et al., 'Survey of sickle-cell disease in England and Wales', *British Medical Journal*, 283 (1981), 1519–21; B. Modell, 'Effect of fetal diagnostic testing on birth rate of thalassaemia major in Britain', *The Lancet*, (15 December 1984), 1383–6.

31 J. Donovan, 'Ethnicity and health: a research review', *Social Science and Medicine*, 19 (1984), 663–70.

32 G. Stamatoyannopoulos, 'Problems of screening and counselling in the haemoglobinopathies' in A. G. Motulsky and J. Ebling (eds), *Proceedings of the Fourth International Symposium on Birth Defects* (Vienna: Excerpta Medica, 1974), 268–76.

33 B. Modell, 'Social aspects of prenatal monitoring for genetic disease' in H. Galjaard (ed.), *'The Future of Prenatal Diagnosis* (Edinburgh: Churchill Livingstone, 1982), 144–59.

34 C. R. Scriver et al., 'β-thalassaemia disease prevention: genetic medicine applied', *American Journal of Human Genetics*, 36 (1984), 1024–38.

35 Modell, op. cit. (see note 30 above).

Chapter 9 New forms of genetic intervention

1 D. R. Parks and L. A. Herzenberg, 'FACS: theory, experimental optimisation and applications in lymphoid cell biology', *Methods in Enzymology*, 108 (1984), 197–241.

2 B. Modell, 'Chorionic villus sampling: evaluating safety and efficacy', *The Lancet* (30 March 1985), 737–40.

3 Anshan Department of Obstetrics and Gynecology, 'Fetal sex prediction by sex chromatin of chorionic villi cells during early pregnancy', *Chinese Medical Journal* 1 (1975), 117–26.

4 Modell, op. cit. (see note 2 above).

5 T. B. Perry et al., 'Chorionic villi sampling: clinical experience, immediate complications, and patient attitudes', *American Journal of Obstetrics and Gynecology*, 151 (1985), 161–66.

6 Modell, op cit. (see note 2 above).

7 Ibid.

8 J. Burn, 'Clinical genetics', *British Medical Journal* (8 October 1983), 999–1000.

9   A. de la Chapelle, 'Human genes: mapping hereditary disorders', *Nature*, 317 (10 October 1985), 472–3.

10 R. White et al., 'Construction of linkage maps with DNA markers for human chromosomes;, *Nature*, 313 (10 January 1985), 101–5.

11 J. Gusella et al., 'A polymorphic DNA marker genetically linked to Huntington's disease', *Nature*, 306 (17 November 1983), 234–8.

12 C. R. Cantor, 'Huntington's Disease: charting the path to the gene', *Nature*, 308 (29 March 1984), 404–5.

13 P. S. Harper and M. Sarfarazi, 'Genetic prediction and family structure in Huntington's Chorea', *British Medical Journal* 290 (29 June 1985), 1929–31.

14 S. Thomas, 'Ethics of a predictive test for Huntington's Chorea', *Nature*, 284 (8 May 1982), 1383–5.

15 P. S. Harper et al., 'Decline in predicted incidence of Huntington's Chorea associated with systematic genetic counselling and family support', *The Lancet* (22 August 1981), 411–13; C. O. Carter and K. Evans, 'Counselling and Huntington's Chorea', *The Lancet* (September 1979)

16 Thomas, op. cit. (see note 14 above).

17 N. S. Wexler, 'Huntington's disease and other late onset genetic disorders' in A. E. H. Emery and I. Pullen (eds), *Psychological Aspects of Genetic Counselling* (London: Academic Press, 1984), 125–46.

18 D. C. Watt et al., 'Probes in Huntington's Chorea', *Nature*, 320 (6 March 1986), 21; J. Gusella, ibid., 21–2.

19 A. Jeffreys et al., 'Hypervariable "minisatellite" regions in human DNA', *Nature*, 314 (7 March 1985), 67–72.

20 A. Jeffreys et al., 'Positive identification of an immigration test-case using human DNA fingerprints', *Nature*, 317 (31 October 1985), 818–19.

21 D. J. Weatherall, 'Prenatal diagnosis of thalassaemia', *British Medical Journal*, 288 (1984), 1321–2.

22 G. E. Seidel, 'Production of genetically identical sets of mammals: cloning?' *Journal of Experimental Zoology*, 228 (1983), 347–54.

23 H. M. Michelmann and L. Mettler, 'Cytogenetic investigations on human oocytes and early human embryonic stages', *Fertility and Sterility*, 43 (1985), 320–2; R. H. Martin, 'Direct chromosomal analysis of the human spermatozoa', *American Journal of Human Genetics*, 34 (1982), 459.

24 T. Beardsley, 'Embryo sexing: cattle now, people next?' *Nature*, 304 (1983), 301.

25 Interestingly the same considerations apply to ovum donation from someone known not to be a carrier of the relevant gene. That would obviate the need for antenatal or pre-implantation diagnosis altogether, but the success rate would again be around 10 to 20 per cent. The same is true of embryo donation.

26 I. Bianco et al., 'Prevention of thalassaemia major in Latium (Italy), *The Lancet* (19 October 1985), 888–9.

27 S. Zeesman et al., 'A private view of heterozygosity: Eight-year follow-up study on carriers of the Tay-Sachs gene detected by high school screening in Montreal', *American Journal of Medical Genetics*, 18 (1984), 769–74.

28 E. de H. Lobo and D. S. Thompson, 'Screening and counselling of school leavers for the carrier state of hereditary anaemias', *British Medical Journal*, 284 (1 May 1982), 1301.

29 M. G. Michaelson, 'Sickle Cell Anaemia: An "interesting pathology" '. *Ramparts* (October 1971), 52–8.

30 R. Hubbard and M. S. Henifin, 'Genetic screening of prospective parents and workers', *International Journal of Health Services*, 15 (1985), 231–51; J. H. Turner et al., 'Legal and social issues in medical genetics', *American Journal of Obstetrics and Gynecology*, 134 (1979), 83–99.

31 M. R. Farfel and N. A. Holtzman, 'Education, consent and counselling in sickle cell screening programs: report of a survey', *American Journal of Public Health*, 74 (1984), 373–5; P. Reilly, 'Government support of genetic services', *Social Biology*, 25 (1978) 23–37.

32 U. Prashar et al., *Sickle Cell Anaemia – Who Cares? A survey of screening and counselling facilities in England* (London: The Runnymede Trust, 1985).

33 L. Kopelman, 'Genetic screening in newborns: voluntary or compulsory?' *Perspectives in Biology and Medicine* (Autumn 1978), 83–4.

34 J. Green, 'Media sensationalism and science: the case of the criminal chromosome' in T. Shinn and R. D. Whitley (eds), *Expository Science: Forms and functions of popularisation* (Dordrecht: Reidel, 1985), 139–61.

35 T. H. Murray, 'Genetic testing at work: how should it be used?' *Technology Review*, 88 (1985), 50–9.

36 J. Green, 'Detecting the hypersusceptible worker: genetics and politics in industrial medicine', *International Journal of Health Services*, 13 (1983), 247–64.

37 Office of Technology Assessment, *The Role of Genetic Testing in the Prevention of Occupational Disease* (Washington; OTA, 1984).

38 M. Lappé, 'Ethical issues in testing for differential sensitivity to occupational hazards', *Journal of Occupational Medicine*, 25 (1983), 797–808.

39 B. Merz, 'Nobelists take genetics from bench to bedside', *Journal of the American Medical Association*, 254 (1985), 3161.

40 Z. Harsanyi and R. Hutton, *Genetic Prophecy: Beyond the double helix* (London: Granada, 1982).

41 J. Stewart, 'Scientific findings that look awkward for socialists: How are we to respond?', *Radical Science Journal*, 8 (1979), 121–3; J. Stewart,

'Schizophrenia: the systematic construction of genetic models', *American Journal of Human Genetics*, 32 (1980), 47–54.

Chapter 10 Gene therapy

1 Unnamed scientist quoted in B. J. Culliton, 'Gene therapy: research in public', *Science*, 227 (1 February 1985), 493–6.

2 W. F. Anderson, 'Human gene therapy: scientific and ethical considerations', *Recombinant DNA Technical Bulletin*, 8 (June 1985), 55–63.

3 A. Flavell, 'Genetic engineering: Drosophila takes off', *Nature*, 305 (8 September 1984), 96.

4 J. R. Beckwith, 'Gene expression in bacteria and some concerns about the misuse of science', *Bacteriological Reviews*, 34 (September 1970), 222–7; J. Beckwith, 'The scientist in opposition in the United States' in W. Fuller (ed.), *The Social Impact of Modern Biology* (London: RKP, 1970), 225–32; see also J. R. Beckwith, 'Social and political uses of genetics in the United States: Past and present', *Annals of the New York Academy of Sciences*, 265 (1976), 46–58.

5 T. Friedmann, and R. Roblin, 'Gene therapy for human genetic disease', *Science*, 175 (3 March 1972), 949–55.

6 S. Rogers, 'Skills for genetic engineers', *New Scientist*, (29 January 1970), 194–6.

7 W. F. Anderson and J. C. Fletcher, 'Gene therapy in human beings: when is it ethical to begin', *New England Journal of Medicine*, 303 (27 November 1980), 1293–7; see also in the same issue K. E. Mercola and M. J. Cline, 'The potentials of inserting new genetic information', ibid., 1297–1300.

8 'Furore over human genetic engineering', *New Scientist* (16 October 1980), 140; N. Wade, 'Gene therapy caught in more entanglements', *Science*, 212 (3 April 1981), 24–5.

9 B. Williamson, 'Reintroduction of genetically transformed bone marrow cells into mice', *Nature*, 284 (3 April 1980), 397.

10 S. J. Reiser et al. (eds), *Ethics in Medicine: Historical perspectives and contemporary concerns* (Cambridge Mass.: MIT Press, 1977).

11 I. E. Thompson et al., 'Research ethical committees in Scotland', *British Medical Journal*, 282 (1981), 718–20.

12 D. Dickson, 'NIH censure for Dr. Martin Cline', *Nature*, 291 (4 June 1981), 369.

13 B. Williamson, 'Gene therapy', *Nature*, 298 (29 July 1982), 416–18.

14 President's Commission for the Study of Ethical Problems in Medicine and Biomedical and Behavioral Research, *Splicing Life: the social and ethical issues of genetic engineering with human beings* (Washington: USGPO, 1982); US Congress, House Committee on Science and

Technology, Subcommittee on Investigations and Oversight, *Human Genetic Engineering* (Hearings, 16, 17, 18 November 1982) (Washington: USGPO, 1982); J. A. Johnson, 'Human gene therapy' (Library of Congress, Congressional Research Service, Issue Brief IB84119, 8 November 1984); Office of Technology Assessment, *Human Gene Therapy* (Washington: USGPO, 1984); RAC Working Group on Human Gene Therapy, 'Human Gene Therapy: Points to consider in the design and submission of human somatic cell gene therapy protocols', *Recombinant DNA Technical Bulletin*, 8 (June 1985), 66–72; Revised version in *Recombinant DNA Technical Bulletin*, 8 (September 1985), 116–22.

15 'Gene im Griff? Eine Zwischenbalanz zur Gentechnik-Kommission', *Wechselwirkung*, 7 (May 1985), 47–51.

16 W. F. Anderson, 'Prospects for human gene therapy', *Science*, 226 (26 October 1984), 401–9; J. A. Griffin, 'Recombinant DNA - Potential for gene therapy', *American Journal of the Medical Sciences*, 289 (March 1985), 98–106.

17 K. Simmons, 'Clinical stage draws nearer in ongoing studies of gene therapy', *Journal of the American Medical Association* 253 (4 January 1985), 13–16; S. Budiansky, 'Gene therapy: US clinical trials imminent', *Nature*, 312 (29 November 1984), 393; G. Vines, 'New tools to treat genetic disease', *New Scientist* (13 March 1986), 40–2; L. Walters, 'The ethics of human gene therapy', *Nature*, 320 (20 March 1986), 225–7.

18 J. Rifkin, *Algeny*, (New York: Viking, 1983); see also S. Budiansky 'Gene manipulation: churches against germ changes', *Nature*, 303 (16 June 1983), 563.

19 Anderson, op. cit. (note 16 above).

20 R. A. Hock and A. D. Miller, 'Retrovirus-mediated transfer and expression of drug resistance genes in human haematopoietic progenitor cells', *Nature*, 320 (20 March 1986), 275–6.

21 M. Robertson, 'Gene therapy: desperate appliances', *Nature*, 320 (20 March 1986), 213–14.

22 See *Points to Consider*, op. cit. (see note 16 above)

23 'Patenting human reproduction', *Institute of Medical Ethics Bulletin* (April 1985), 11–12.

24 R. E. Hammer et al., 'Production of transgenic rabbits, sheep and pigs by microinjection', *Nature*, 315 (20 June 1985), 680–3.

25 S. P. Leibo, 'The embryo transfer industry: profile and prospects for the future' in *Biotechnology: High Technology Industries: Profiles and Outlooks* (Washington: US Department of Commerce, 1984).

26 Anderson, op. cit. (see note 16 above).

27 G. F. Stranzinger, 'Problems of genetic engineering in animal breeding' in W. Arber et al., *Genetic Manipulation: impact on man and society* (Cambridge: Cambridge University Press, 1984), 235–8.

28   J. A. Miller, 'Interspecific chimerism: sheepish goats and goatish sheep',
      BioScience, 34 (May 1984), 290.
29   Williamson, op. cit. (see note 13 above).
30   J. Glover, *What Sort of People Should There Be?* (Harmondsworth:
      Penguin, 1984).

Chapter 11 Back to nature

1   Congress: 'Women against gene technology and reproductive tech-
     nologies', Bonn, West Germany, 19–21 April 1985, 3 – page photo-
     copied conference resolution.
2   B. A. Santamaria, 'By whom begot?' *Salisbury Review* (January 1985),
     11–18. Quotation from p. 13. I am grateful to Geoffrey Price for
     bringing this to my attention.
3   D. Holbrook, 'Medical ethics and the potentialities of the living being',
     British Medical Journal, 291 (17 August 1985), 459–62.
4   *Report of the Committee of Inquiry into Human Fertilisation and
     Embryology* (London: HMSO, 1984: Cmnd 9314); M. Warnock, *A
     Question of Life* (Oxford: Blackwell, 1985).
5   Warnock, op. cit., pp. ix–x.
6   H. Brody, 'Autonomy revisited: progress in medical ethics: discussion
     paper', *Journal of the Royal Society of Medicine*, 78 (May 1985), 380–7.
     The conception of autonomy discussed in the text is taken from the work
     of B. L. Miller, *Hastings Center Report*, 11 (4), (1981), 22–8.

Index